JEAN WALRAND

I0390691

UNCERTAINTY

A User Guide

AMAZON

o

PUBLISHED BY AMAZON

First printing, April 2019

Contents

Dedicated to my students and colleagues who taught me what I possibly know about uncertainty, and to Melanie

Acknowledgments

> *I feel a very unusual sensation – if it is not indigestion, it must be gratitude.*
>
> Benjamin Disraeli

I have been fortunate to be surrounded by amazing colleagues and students.

I learned Applied Probability from the best in the field, including Professors David Aldous, Venkat Anantharam, Vivek Borkar, Terry Fine, Bruce Hajek, P.R. Kumar, Jim Pitman, Kannan Ramchandran, David Tse, John Tsitsiklis, Pravin Varaiya, and Eugene Wong.

I taught courses on related topics with Professors Venkat Anantharam, Abhay Parekh, Shyam Parekh, Satish Rao, and Sanjit Seshia.

I am grateful for the comments on this little book from Professors Venkat Anantharam, Longbo Huang, Abhay Parekh, Balaji Prabhakar, Satish Rao, and Anant Sahai, and my wife Annie. I wish I could blame them for the remaining errors.

This book started with an invited lecture at the Université Alioune Diop de Bambey, Sénégal, organized by Professor Assane Gueye.

Thank you!

Preface

If life were predictable it would cease to be life, and be without flavor.
Eleanor Roosevelt

Indeed, uncertainty is the spice of life.

Of course, uncertainty is scary. We live with the risk of natural disasters, accidents, and diseases. But uncertainty is exciting. We buy a lottery ticket hoping for a big prize, enjoy a new song or book, are fascinated by the news, and thrilled by professional sports. Also, fundamentally, uncertainty drives evolution, and probably our consciousness and illusion of free will.

It is difficult to think clearly about such an omnipresent and essential aspect of our life. We do not assess risks rationally, get confused by statistics, mistake correlation for causation, and find it difficult to make choices with uncertain outcomes.

In this little book, I discuss topics that sharpen our understanding of uncertainty. These topics include interpreting data, investments, games, and machine learning; they are explained with intuitive examples. I also cover some speculations about the role of uncertainty in our consciousness and perception of time.

I hope you will share my fascination with these topics.

Jean Walrand
Berkeley, April 2019

1

What is Uncertainty?

> We sail within a vast sphere, ever drifting in uncertainty, driven from end to end.
>
> Blaise Pascal

Where does uncertainty come from? Can it be reduced or even eliminated? Is it unavoidable? How can one model it? What can one do about it?

Sources of Uncertainty

Uncertainty often results from a combination of lack of knowledge, chaos, and the limited computing power of our brain.

When someone flips a coin, we find it almost impossible to predict the outcome.

We do not know precisely the spin rate ω of the coin nor its vertical velocity v when it leaves that person's hand. The figure shows the outcome of a coin flip ('heads' = H or 'tails' = T) as a function of ω and v. Note that a tiny change in these values changes the outcome, an effect called *chaotic dynamic*, or *chaos* for short. If we knew precisely ω and v and the physical characteristics of the coin and of the surface it falls on,

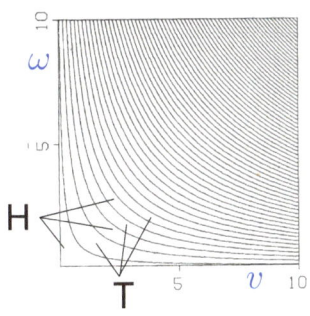

we could use a computer to simulate the coin flip and predict whether it will yield 'heads' or 'tails'.

Of course, our brain would not be able to perform the simulation. Thus, the coin flip would still *appear* unpredictable to us, because of our limited computing power.

Similarly, forecasting the weather is difficult, even with a supercomputer, because of lack of precise knowledge of the current state and of the chaotic dynamic.

When you shuffle a deck of cards, unless you are a magician, you do not know precisely the ordering after one shuffle: does the third card from the half deck in your right hand slide on top or under the fifth card of the half deck in your left hand? This lack of knowledge results in an unpredictable ordering after multiple shuffles.

Every marble in a well shaken bag of identical marbles is equally likely to be picked. This is because the shaking of the marbles is a chaotic process and a small variation in the acceleration of the bag results in a different configuration of the marbles in the bag.

Thus, the world appears unpredictable to us because of our limited ability to compute and to sense, even when its evolution follows laws that have no uncertainty.

 ## Does God Play Dice?

In contrast with Classical Physics, Quantum Mechanics teaches us that the world is fundamentally random. That theory, which is extraordinarily accurate in its predictions, says that one can only compute the likelihood of the behavior of electrons and photons. One cannot predict the trajectories of these particles. They are *irreducibly* unpredictable.

The meaning of this statement is rather subtle. It is qualitatively different from saying that one cannot predict the outcome of a coin flip. In the case of an electron or a photon, it is *impossible* to know exactly their position and velocity. *Heisenberg*'s principle say that the more precisely we know the position, the less precisely we can know the velocity.

Thus, it is possible, in principle, to predict the outcome of a coin flip. However, it is fundamentally impossible to predict the trajectory of an electron or a photon. Those particles do not follow deterministic laws of motion.

This is philosophically disturbing to most people (including me), because our 'everyday' experience of the world is deterministic. When you look at billiards balls moving on the pool table, you realize that the motion is complex but you know that the future is entirely determined by the current positions and velocities of the balls. That is, unless *Shrödinger*'s cat jumps on the table.

Some scientists (notably Albert Einstein, Boris Podolsky, and Nathan Rosen) were hoping that this lack of predictability of quantum mechanics is due to some 'hidden variables' that determine the trajectories.

Remarkably, John Bell proved that there cannot be 'local' hidden variables in quantum mechanics. His argument is as follows. Quantum mechanics predicts that the polarizations of a photon measured at two angles separated by

A	B	C	Matches
0	0	0	3/3
0	0	1	1/3
0	1	0	1/3
0	1	1	1/3
1	0	0	1/3
1	0	1	1/3
1	1	0	1/3
1	1	1	3/3

120 degrees match with probability 1/4. This prediction has been verified experimentally. However, if the polarizations at 0, 120, and 240 degrees were governed by hidden variables A, B, C that take values in $\{0,1\}$, then the average probability of a match between two of these variables would be at least 1/3. For instance, say that the variables A, B, C are equal to 0, 1, 1 as shown in red in the figure. If we pick two directions at 120 degrees, by symmetry they are equally likely to be AB, BC, or CA, since the orientation of the photon is unpredictable. If the polarizations we observe are AB or CA, then they do not match; if they are BC, they do. Hence, the odds are 1/3 that they match. As you can see from

the table, the odds are always at least 1/3, no matter what the values of A, B, C are. This contradicts the 1/4 observations, so the polarizations cannot be governed by hidden variables.

If you do not want to accept this basic indeterminism, you have to accept bizarre non-local variables that communicate their values to each other faster than light. The conceptually simplest view is then that the world is non-deterministic. God plays dice, as Einstein refused to accept. John Bell said "So for me, it is a pity that Einstein's idea doesn't work. The reasonable thing just doesn't work."

Interestingly, Einstein was comfortable with the idea that time is not absolute, an idea that I personally find even more disturbing than the fundamental randomness of the world.

Does the basic randomness of quantum mechanics have 'real-world' effects? Does it change our personal experience, or is the effect confined to unobservable details of the motion of elementary particles?

Macroscopic systems follow laws of motion that are not random, like the motion of balls on a billiards table. This motion is not affected in a measurable way by the quantum uncertainty.

However, many physical systems respond to the behavior of elementary particles and produce results that are fundamentally unpredictable yet visible. For instance, a Geiger counter detects alpha particles, beta particles and gamma rays. The counts that this device

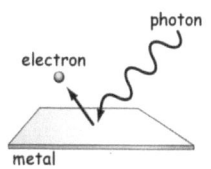

produces are fundamentally unpredictable. Similarly, one can amplify the current that a photon produces when it perturbs and ejects an electron from a metal. That current is then fundamentally unpredictable. Quantum cryptographic systems and quantum computers are other examples of systems that respond to the unpredictable behavior of elementary particles. More basically, our world would not be possible without the discrete nature of energy, as Max Planck explained in his study of *black body radiation*. Indeed, classical physics states that a black body in equilibrium should have the same energy at all possible frequencies of standing waves, because those different modes can exchange energy. Since the number of such frequencies is infinite – for any multiple of the frequency of a standing wave corresponds also to a standing wave – this is impossible. The resolution of this paradox that agrees with experiments is that energy exists in discrete units, called *quanta*, proportional to their frequency. This implies that the body cannot have energy beyond some frequency.

Randomness and Consciousness

If our behavior is (quantum?-) mechanical, like that of a fancy robot, why are we conscious (self-aware, with a 'qualia' experience)? How does consciousness happen? Why is that useful?

I certainly don't know, but let me offer some personal reflexions. One has to distinguish *functional model* and *explanation* of consciousness. By definition, consciousness can only be experienced from 'within'. There is no possible test to determine whether some being other than us is conscious. Thus, questions such as 'is my friend, my dog, my computer conscious' have no verifiable answers. In that sense, the question of consciousness is not a scientific problem: it cannot be tested. Consciousness is what we experience; it is the emergent effect of our nervous system and brain.

One can study the components of the nervous system and the brain and postulate that consciousness requires those components. These hypotheses can be verified by interviewing people with different types of brain injury. However, even if we identified precisely the pieces needed for consciousness, I believe we would never be able to explain how consciousness emerges. It is what it is, because that is what we call it.

One might build some structural model where consciousness emerges by observing the correlation between random actions and their consequences (William James, Henri Poincaré, Daniel

Dennett and many others). For instance, the brain randomly decides to close the right hand. Our senses then observe the effect. In that way, a map gets formed of the effect of decisions. This map tunes our filtering process so that it can now select actions based on their effects. Thus, the brain can decide to close the right hand in order to grasp an object.

Progressively, the mind constructs an abstract layer where it makes selections based on their consequences. As an illustration, consider how my granddaughter Melanie is learning to grab a toy. She first makes random-looking gestures. Some of these gestures touch the toy. Eventually, the brain selects the gestures that have the "desired" consequences.

Combined with memory, this layer may create a sense of 'self', i.e., a sense of existence. This sense of self is very useful because it promotes the instinct of self-preservation and evolution certainly favors animals that have that instinct.

Thus, one may be able to create a functional understanding of the brain and sketch a plausible structure that consists of a random generator of possible actions, a filter that gets tuned over time by watching consequences, a memory that stores a further layer of abstraction where 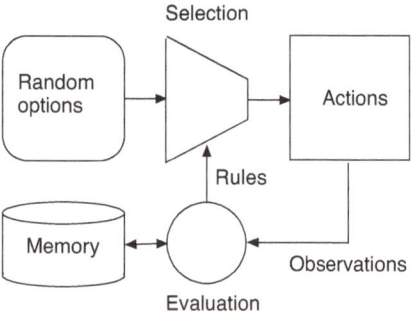 some events produce special chemicals that are favored. In such a model, consciousness is the result of a feedback loop between actions and consequences. However, this structure cannot explain the 'sensation' of awareness, pleasure, pain, happiness, distress, and I believe nothing can.

Along the same line of pointless speculation, one may wonder about the consciousness of animals. It may be that in very simple animals, say cockroaches, consciousness is inexistent. For a complex animal like a dog, it may be close to that of a human. There probably is a minimum degree of complexity required for consciousness: a way to capture the correlation between decisions and effects and also to store such correlations. I think that consciousness is a matter of degree and not an all-or-nothing phenomenon. Consciousness is sharpened by a practice of mindfulness and diluted while being sleepy or under the influence of some drugs (so I am told).

Could a computer be conscious? The computer would not have a body awareness without all the senses that provide feedback on actions. The 'wet stuff' we are made of, or a perfect emulation of it (e.g., movie *The Matrix*), is necessary for a human awareness of existence. Possibly, the computer  does not have the minimal complexity and architecture that trigger an elementary awareness. Also, if one eventually reaches that complexity, its sense of self-awareness might emerge but would so different from a human's that we have no clue as to what it would 'feel like' to be a computer, much as we do not know what it is like to be a bat (Thomas Nagel). We will never know because we will never be 'within' that entity.

Uncertainty and Free Will

Whereas Williams James and Henri Poincaré believed that the selection among random options that the brain performs is free (*deliberate*), Dennett and others think it is *determined*, but possibly somewhat randomly, by prior experience.

After all, what else would govern the choice process? What immaterial magic could possibly be in the 'gap' between options and choice? How would this immaterial thing act on the nervous system? Let's face it: that just doesn't make sense. Even

if the selection is determined by the past, through the complex state of the brain, after this process we are left with the *illusion* that we made a choice. In fact, experiments seem to show that the decision is committed before the person becomes aware of it (e.g., Kerri Smith). Let us note here that this question remains controversial among researchers, possibly because it is difficult to accept that free will is just an illusion.

The random appearance of thoughts that pop up in our head may serve the purpose of exploring a range of possibilities, in the same way that the random motion of a robotic vacuum cleaner or lawn mower helps those devices scan a fairly arbitrary area without having to be programmed for that particular environment.

If we could observe all the mechanisms of our neurons, we would see that our next "choice" is a consequence, possibly random, of the past. The illusion of free will would vanish and be replaced by the awareness of the unfolding of a complex sequence of firings of our neurons governed by

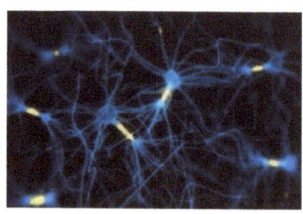

the laws of physics. In that sense, the illusion of free will, like the illusion of the randomness of a coin flip, is due to complexity that overwhelms our cognitive abilities.

Invoking the fundamental quantum uncertainty here (e.g., Roger Penrose, Matthew Fisher, John Searle) may replace the mechanical determinism that some people find objectionable by a fundamental quantum randomness, but I do not see how this can be comforting.

Is a fundamentally random quantum-mechanical selection more free than a selection driven by a deterministic but pseudo-random generator, like a coin flip? Does a photon choose whether it goes through a semi-transparent material because the outcome is fundamentally random? With or without quantum randomness, the mind is

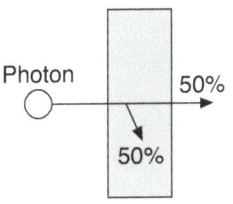

Semi-transparent
material

a machine whose sequence of steps is dictated by the laws of physics that leave no room for free choice.

In any case, quantum effects are not necessary for us to experience randomness. The apparent randomness of a coin flip does not involve quantum effects. So, I don't quite see how such effects would change our experience. The hypothesis seems unnecessary.

When I reflect upon my own choices, I can almost always explain why I made them. Invariably, it is because of previous experiences. My choices are complex consequences of the past. I am willing to accept the idea that I have no free will and that my consciousness, the tip of the huge hidden iceberg of neural activities, is only a spectator of the sequence of events.

I find irrelevant the usual objection to this mechanistic view: 'What about personal responsibility?' It is irrelevant because the merits of a conjecture do not depend on its agreement with unquestioned prior beliefs.

Wishful thinking is not a reliable method for discovering truth. I was brought up by parents and educators who impressed upon me a sense of ethics and morals. My decisions tend to follow those principles, inevitably but probably with an unavoidable dose of randomness and imperfection.

Uncertainty and Evolution

When looking back at the apparition of life and the evolution of plants and animals, it is impossible not to be amazed by the role of randomness. The basic drivers of evolution are random mutations and selections. Errors in copies of genes produce mutations. Sexual reproduction is another major source of randomness and the diversity that it induces is probably beneficial for speeding up evolution.

As a concrete example of the role of randomness, the survival advantage of homo sapiens may be due to a change in the climate in Africa. What were the odds of that? Of course, even if all this was extremely unlikely, given that we are here, it certainly happened.

The tree of life produces random branches and those well matched to the environment and the competition survive. Life is a remarkable game of chance. This random exploration of variations results in a rich multiplicity of life forms that automati- 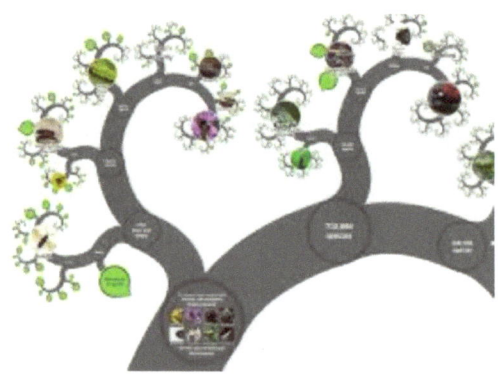 cally get selected for a changing environment. Of course, evolution has limitations. We are currently living through a massive extinction of plant and animal species. The environment is rapidly becoming toxic for many life forms. Humans are remarkably clever at controlling and dominating their circumstances. Nevertheless, it may be inevitable that we will soon appear on the list of endangered species, specially as we don't seem to take climate change and pollution seriously enough to modify our lifestyle.

Modeling Uncertainty

Enough, and possibly too much, of these pointless unscientific speculations. Let us come back to the main theme of this book: uncertainty. How do we model this concept? Fortunately, many bright thinkers from Jacob Bernoulli to Andrey Kolmogorov developed a clear and simple formalism.

Imagine a bag with 100 marbles that are identical, except for their color. There are 27 red marbles and the others are blue. You shake the bag and pick up a marble without looking. The odds that you pick a red marble are 27/100. We call these odds the *probability* of picking a red marble. We write $P(red) = 0.27$.

What does this really mean? A simple interpretation is that if you were to perform this experiment a large number of times, the fraction of those experiments when you pick a red marble (and put it back) would be about 27%. This consequence, called the *Law of Large Numbers*, is not obvious, but is verified in experiments. It can be proved mathematically by introducing the axioms of Kolmogorov (including countable additivity).

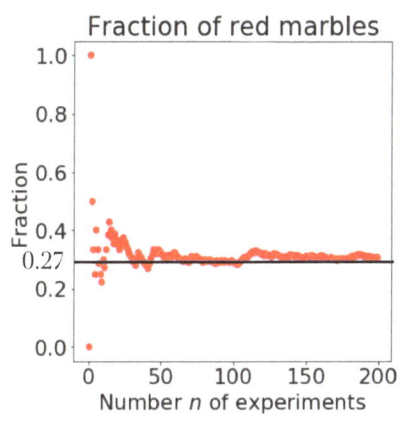

The fact that this mathematical model, with those axioms, yields results that agree with experiments indicates that the model is probably able to predict other results that are relevant in the real world. In this sense, Probability Theory is somewhat like Physics. In Physics, one chooses laws that are not invalidated by experiments. Similarly, in Probability Theory, one chooses axioms so that predictions are not invalidated by experiments.

Thus, although mathematical theories are self-contained and independent of the physical world, the goal of Probability Theory is to explain real world observations. One can construct probability theories where the Law of Large Numbers does not hold and whose conclusions do not match experimental results. Kolmogorov's model seems to be appropriate.

Let us note that some philosophers object to the axiom of countable additivity. That axiom implies that it is not possible to assign an equal positive probability to all the elements of an infinite set, such as the integers $\{1, 2, 3, \ldots\}$. Thus, a lottery where all these integers have the same probability of being selected cannot be defined. I don't find this fact to be a reason to reject countable additivity. No one could offer such a lottery anyway and no one should trust a charlatan who pretends he does.

If a coin is loaded (i.e., biased, or unfair), the odds that a flip yields 'heads' are not 50%. Instead, the fraction of 'heads' in a large number of flips may be 27%. In this case, one calls 27% the probability of 'heads' and we write $P(heads) = 0.27$.

Now assume that you get A dollars when the coin flip produces 'heads' and B dollars when it produces 'tails'. If you perform this experiment 100 times, then about 27 times you get A and the other 73 times you get B. The total amount you expect to get after 100 experiments is then $27 \times A + 73 \times B$. Accordingly, the average amount you expect to get per experiment is

$$\frac{27 \times A + 73 \times B}{100} = 0.27 \times A + 0.73 \times B.$$

We call this quantity the *expected value* that you get after one experiment. Thus, the interpretation of the expected value is the average value you would expect to get per experiment, if you were to perform the experiment many times.

Let us write $P(A) = 0.27$ to indicate that the probability that you get A is 0.27. Similarly, we write $P(B) = 0.73$. Then we can write that the expected amount you get per

0.27 A

0.73 B

Expected reward:
0.27A + 0.73B

experiment is
$$P(A) \times A + P(B) \times B.$$

If there were more than two possible outcomes, say $A, B, \ldots, K$, then the expected amount per experiment would be

$$P(A) \times A + P(B) \times B + \cdots + P(K) \times K.$$

Thus, the *expected value* of some random quantity (e.g., the value that you get after an experiment) is the sum over all its possible values (e.g., $A, B, \ldots, K$) of the product of the probability of that value (e.g., $P(A), P(B), \ldots, P(K)$) times the value itself. We will use this definition a few times in the book.

This simple model of *probability* and *expected value* summarizes the uncertainty. It abstracts away the details of the physical experiment that produces the outcome. The model is the same whether you pick a marble from a bag or you flip a loaded coin. This powerful abstraction is a great simplification that helps us think about uncertainty.

 ## What Can One Do about Uncertainty?

Uncertainty is omnipresent in our life. What can one do about it? This is precisely the question that we explore in this little book. We will examine how to think more clearly about uncertain events and how to make choices that have uncertain consequences.

2

Risks

> *Take calculated risks. That is quite different from being rash.*
>   George S. Patton

Do you get a flu shot? Are your kids vaccinated? Do you have a house earthquake insurance? Do you buy a two-year warranty extension for your new smartphone? Are you afraid to fly? How about to sky-dive? Do you ever buy lottery tickets? What about used cars? Do you wear a helmet when you bike? Do you downhill-ski? Are you more afraid of a terrorist attack or of driving to work? Would you leave a steady job to join a new startup?

We are constantly facing such questions and our answers tend to be inconsistent.

Likelihoods

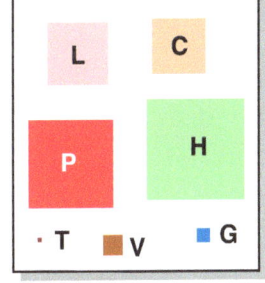

Smoking is about two thousand times more likely to kill us than terrorism. (In the figure, the area of each square is proportional to the corresponding lifetime probability of death for a U.S. male.) Yet, some smokers are afraid of terrorism. Could it be that smoking provides a satisfaction that justifies the high danger of lung cancer? Or is it the absence of control in a terrorist attack? Or, as I believe,

C colon cancer
G gun shot
H heart disease
L lung cancer
P prostate cancer
T terrorism
V vehicle accident

a poor awareness of the risks.

In a curious coincidence, food poisoning kills about 65 times more people in the US than terrorist attacks; yet, the Department of Homeland Security has about 65 times more employees than there are food inspectors. Is this acceptable? (About 5,000 deaths per year in the US from food poisoning vs. 76 from terrorism; DHS had 229,000 employees and there were about 9,200 employees for food inspection.)

Obesity is ten times more deadly than car accidents (see the figure). Yet, public health programs to reduce obesity are insignificant compared to the cost of seat belts, air bags, traffic policing, and other measures concerning road safety. Some will argue that we should be allowed to take chances with our own lives as long as we do not endanger others. However, the premiums for health insurance reflect the poor choices of many. Is it acceptable to ask others to pay for our choices?

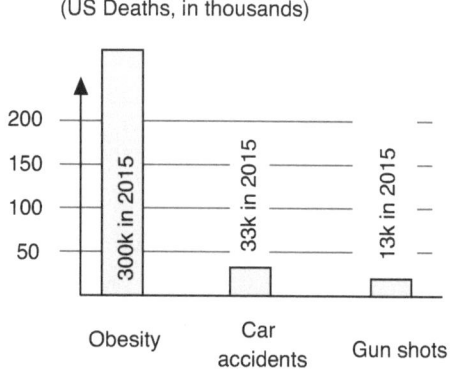

(US Deaths, in thousands)

You are many times more likely to die in a car crash than because of an earthquake, even if you live along the Hayward fault in Berkeley, California. Is your fear proportional to the risk?

Before you board a plane, you go through a TSA checkpoint that is supposed to screen out explosives and weapons in carry-on luggage. Repeated tests show that these checkpoints are 95% ineffective (according to ABC News). How about the larger number of checked luggages that may not be properly screened. Should we be worried or upset, or both?

(Worldwide, about 5,800 people died because of airplane terrorism since 1940, including on 9/11/2001, according to a CNN report of 12/2017.)

So What?

You get the point. We are inconsistent in our appreciation of risks and our efforts to mitigate them. What can we do about this inconsistency? I think that an awareness of relative risks is useful. For instance, parents who refuse to get their kids vaccinated should find out about the real risks and costs. People who ingest food supplements in the belief that they are generally useful should get better informed. If you are afraid of Diet Coke™ or of living close to a cell phone tower, are your fears based on data or on some fuzzy belief?

It is difficult to make sense of statistics. Often, we accept ideas without thinking about them. That may be fine. However, once in a while, it may be a good idea to take a second look, particularly if these statistics may inform important decisions.

3
Causation

cum hoc ergo propter hoc

Latin for "with this, therefore because of this."

Does eating ice cream cause sunburns? Obviously not. However, if
we look at 1,000 people who had
some ice cream in the last week and
1,000 who did not, we almost certainly find more people with sun-
burns
burns in the first group than in the second.

This example shows that one should not confuse 'correlation'
and 'causation'. Events are said to be (positively) correlated if
they tend to happen more frequently together, like sunburns and
eating ice cream. Causation means that one event tends to cause
another.

In our example, being in hot
weather, such as on vacation in
Hawaii, tends to cause sunburns and
also a desire for ice cream. Thus, eat-
ing ice cream and getting sunburned
are more likely to happen together
because they have a common cause,
even though one does not cause the other.

I like this example because it is particularly clear. However,
many situations are much less obvious. It is easy to observe correlation, but causation is generally much more difficult to establish.

For instance, smoking and lung cancer are strongly correlated. Now, does smoking cause cancer or does cancer cause smoking? Or is there a third event that causes both? Intuitively, we know that it is smoking that causes cancer. However, how do you prove it?

One might look at the sequence of events. If John developed lung cancer after smoking, it is probably smoking that caused the cancer, and not the other way around. But, what if there were another event, say depression, that causes both smoking and lung cancer, but happens to cause smoking faster than cancer? Of course, this argument is fanciful, but I hope it makes the case that causation is difficult to establish.

A neighbor swears by vitamin C as a way to avoid getting a common cold. Over the last four years, she has been ingesting megadoses and has not had a cold. My former dentist recommended homeopathic potions for the pain after a tooth extraction. Many of his patients testified that it works: they had little pain when taking such concoctions. A friend consumes garlic to help with her arthritis. She remembers many times when she had a good day after eating garlic.

Is there any truth behind these beliefs? I refused the homeopathic liquids of my dentist and also experienced little pain. My placebo was my confidence that I would not feel pain.

 ## Intervention Studies

A standard method to establish causation is an *intervention study*. To conduct such a study for ice cream and sunburns, one would choose randomly $1,000$ people in a group of $2,000$ and have them eat some ice cream and ask the others not to eat any. After a few days, one would then compare the number of people with sunburns in the two groups. Of course, one hopes that the people who had ice cream did not go sunbathe to warm up!

To make sure that the experimenter does not manipulate the

study to prove his hypothesis, one may use a *double-blind* study. In such a study, the experimenter who is counting the sunburns does not know who ate ice cream. This information is revealed only after the numbers have been collected.

Note that intervention studies cannot be performed in many situations. For instance, would you ask a group of people to drink heavily before driving to study if drinking contributes to accidents?

The important statistical question concerns the size of the group, also called *sample size*. Is 1,000 sufficient? Of course, this depends on the strength of the causation. For instance, say that 52 people have sunburns among the 1,000 who ate ice cream while only 42 do among those who did not eat ice cream. Is this difference sufficient to conclude that eating ice cream causes sunburns?

Fluctuations are unavoidable with small samples. It is not surprising that medical studies contradict one another every few years! Today, a study may declare that drinking coffee is linked to an increase in some specific cancer. Two years from now, another study will claim the opposite.

In the next section, we explore the question of *statistical significance*.

 ## Flipping Coins

You have a coin. The coin is loaded (i.e., biased, or unfair): the odds that it yields 'heads' are p. For instance, if $p = 33\%$, it yields 'heads' 33% of the time, in the long run.

However, you do not know p and you try to estimate it by flipping the coin 1,000 times. Assume that F is the fraction of 'heads' among these 1,000 flips. You expect F to be close to p. However, you suspect that F may not be exactly equal to p. After all, you might have been lucky and get an unusually large number of 'heads'.

The key fact is that F is very likely not to differ from p by more than 3%. Thus, if $F = 31\%$, then it is very likely that p is between $F - 3\% = 28\%$ and $F + 3\% = 34\%$. By very likely, one actually means with 95% chance. That is, if you were to repeat the experiment one hundred times, then about 95 of those times the fraction F of 'heads' would not differ from p by more than 3%.

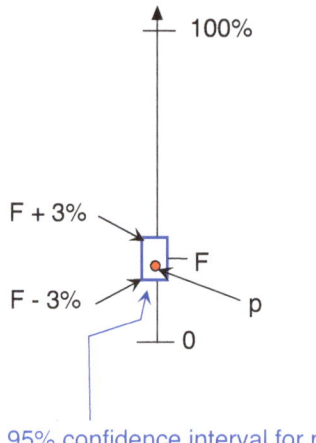

95% confidence interval for p

One says that $[F - 3\%, F + 3\%]$ is a 95%-*confidence interval* for p. This is a mouthful, but it simply says that we can be 95% confident that the actual value of p is in the interval $[F - 3\%, F + 3\%]$.

Let us look at two typical applications of this simple fact.

 Public Opinion Poll

You interview $1,000$ people selected at random in the U.S. and you ask them whether they would vote for Michelle Obama for President in 2020. You find that 62% of these people say yes. Then we know that there is a 95% chance that the fraction p of the voters in the general population who would vote for Michelle is between $62\% - 3\% = 59\%$ and $62\% + 3\% = 65\%$. At least, if they were to vote today.

Experts in polling will point out that our discussion is a bit simplistic. It assumes that all the voters behave like flips of the same coin. In fact, the population is fragmented. We know that men and women vote differently, and so do democrats and republicans. Thus, our analysis should be refined. If you do take all these effects into account, you find that the confidence interval should be broader than F plus minus 3%, possibly F plus minus 6%. Still good odds for Michelle.

Effectiveness

Let's revisit our ice cream and sunburns example.

Using our basic fact about coin flips, we find that the odds that someone who ate ice cream gets a sunburn are between 52/1000 minus 3% and 52/1000 plus 3%, i.e., between 2.2% and 8.2%. Similarly, for people who did not eat ice cream, it is between 42/1000 minus 3% and 42/1000 plus 3%, i.e., between 1.2% and 7.2%.

Thus, these numbers are inconclusive: they are not statistically significant. The difference could be due to
chance, not to an effect of ice cream.
Such inconclusive results are often

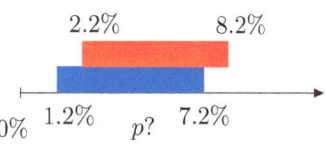

misinterpreted as proving that eating ice cream does not cause sunburns. This is incorrect. The confidence intervals say that the study is inconclusive. The study does not prove nor disprove the causation. The sample size is too small. The results of this study should not be published.

Assume there are two studies. The first one involved 1000 people who smoked for twenty years and 210 of them developed lung cancer, and another 1000 people who never smoked and 180 of them developed lung cancer. The second study involved one hundred thousand people who smoked for twenty years and 20,534 of them developed lung cancer, and another one hundred thousand people who never smoked and 18,147 of them developed lung cancer.

For the first study, the 95% confidence interval for the probability of cancer among smokers is $21 \pm 3\% = [18\%, 24\%]$. For the nonsmokers, it is $18\% \pm 3\% = [15\%, 21\%]$. This study is inconclusive since these two confidence intervals overlap. The incorrect conclusion is that this study proves that smoking does not increase the odds of developing lung cancer. The correct conclusion is that this study is inconclusive: it neither proves nor disproves anything and its results should not be published

For the second study, the confidence intervals for nonsmokers and smokers are $18.147\% \pm 0.3\% = [17.8\%, 18.5\%]$ and $20.53\% \pm 0.3\% = [20.2\%, 20.9\%]$, respectively. (In Chapter *Comments*, we explain that if the sample size is 100 larger, the width of the confidence interval is ten times smaller. Thus, for a sample size of $100,000$, the width is $\pm 0.3\%$ since it was $\pm 3\%$ for a sample size of 1000.) This study shows that, with reasonable confidence (95%), smoking increases the odds of developing lung cancer by at least 1.7%, from at most 18.5% to at least 20.2%. Indeed, there is only a 2.5% chance (half of 5%) that a nonsmoker has a probability larger than 18.5% of developing lung cancer and only a 2.5% chance that a smoker has a probability less than 20.2% of developing lung cancer. Accordingly, there is only a 5% chance that the difference between the odds of a smoker are less than 1.7% more than those of a nonsmoker to develop lung cancer. This little analysis is not too tricky, but it must be done carefully.

It is important to understand that these studies do not contradict each other. The first one was inconclusive; the second one is quite clear. Serious studies should not contradict one another. If they do, someone made a mistake, either in collecting the data or in the analysis.

Too frequently, results of inconclusive studies are published without statistical analysis. As a consequence, studies with conflicting results frequently appear in the media. This practice should stop.

Simpson's Paradox

It is so easy to be misled by statistics. Here is an example known as Simpson's paradox.

Imagine the situation summarized in the figure.

A university has two colleges A and B. Nine hundred female students apply to college A and 90 are admitted. One hundred female students apply to college B and 30 are admitted. Over-

	College A	College B	Total
F	90 / 900	30 / 100	120 / 1000
M	5 / 100	180 / 900	185 / 1000

all, the university admitted 120 of the 1000 female students who applied.

The numbers for male students are different. One hundred of them applied to college A and 5 were admitted; also, nine hundred male students applied to college B and 180 were admitted. Overall, the university admitted 185 of the 1000 male students who applied.

The overall statistics suggest that the university discriminates in favor of the male students since it admits a much larger fraction of male than female students: 18.5% of male students vs. 12% of female students However, both colleges admitted a larger fraction of female students than of male students. Indeed, college A admitted 10% of the female students and only 5% of the male students; college B admitted 30% of the female students and only 20% of the male students. One might accuse the colleges of discriminating in favor of female students.

What is happening here is that female students tend to apply to college A that is much more selective than college B, whereas the male students tend to apply to the less selective college B. Thus, it is not the university that discriminates against female students. It so happens that the female students choose a more selective college and therefore a smaller fraction of them are admitted.

 Hidden Facts

When looking at overall statistics, one is led to think that the university discriminates against female students. The college statistics tend to show that they discriminate against male students.

Now, imagine that the colleges are made up of departments and that the students in fact apply to departments and not to colleges. It might be that if we look at the department numbers, we are led to the conclusion that they discriminate against female students . . ..

By choosing the statistics that we reveal, we can promote a specific conclusion. It is not difficult to lie with statistics!

 Tax Cut

Imagine a hypothetical country with one million rich people who make $1M per year and nine million poor people who make $50k per year, all before tax. To reduce the income disparity, the government decides to cut the taxes of the poor people by 10% and the taxes of the rich people, but only by 5%.

The differences in the reductions of the tax rates seems to favor the poor people. However, together, the rich people get to keep $50B more together while the poor people get only to keep $45B more together. So more than 50% of the tax cut go to the rich people. Is this fair?

Again, by choosing which statistics you reveal, you can change the conclusion.

 Selecting Studies

An even much cruder form of deception is frequent in the media and is illustrated by the following elementary example. Imagine that Bob and Alice each flip the same coin ten times. Bob gets 6 tails and Alice gets 3. Charles wants to promote the idea that the coin is biased in favor of tails. He publishes the results of Bob's experiment and ignores those of Alice.

 ## Cautionary Tale

The examples in this chapter should serve as a cautionary tale.

Next time you hear of a study on the effectiveness of vitamin C at preventing a cold, or the impact of coffee on high blood pressure, or of the effect of a child spending too much time in front of a screen on his attention deficit, use the basic facts about coin flips to check the significance of the study.

4
Inference

You don't feel well and go to your doctor. After a few tests, the doctor makes a diagnosis. She thinks that you probably have some form of food poisoning.

The doctor faces a difficult problem: given the symptoms, what is the most likely cause? Also, is that conclusion un-ambiguous or should the doctor order more tests?

Interestingly, there is a wrong way and a correct way to think about such problems. You might say that this is true about all problems, but in this particu-lar case it turns out that the wrong way is often the one we think about. Let us look at a simple example to clarify the issues.

Spades --> 100% Ace
Hearts --> 50% Ace

Ace --> 60% Spades

Hearts and Spades

We have seven cards as shown in the figure: three aces of spades, two aces of hearts, and two 8 of hearts. You shuffle the cards face down, close your eyes, select one, and you are told it is an ace.

What are the odds it is a spades?

The wrong way to think about this question is to say that your card is very likely to be spades because all the spades are aces but only half of the hearts are aces.

The correct argument is of course that three out of five aces are spades. Thus, the odds that your ace is spades are 60%.

 Flu or Ebola

Imagine you have a high fever. Do you have the flu or Ebola?

The wrong way to think is that you probably have Ebola since everyone with Ebola has a high fever, but only a fraction of people with the flu have such a fever.

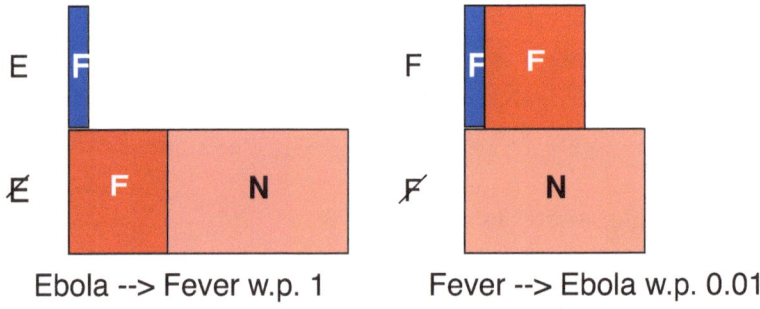

Ebola --> Fever w.p. 1   Fever --> Ebola w.p. 0.01

The correct analysis is to estimate the fraction of people with Ebola among the feverish population. This fraction is very small, because there are so few people with Ebola, so you most likely do not have Ebola. Thomas Bayes is credited with clarifying how to do such calculations.

Tuning A Car Alarm

You are designing car alarms that use motion and noise as indicators of a burglary. You have to tune the sensitivity of the alarm. If it is too sensitive, it causes many false alarms. If it is not sensitive enough, it fails to detect some burglaries. What is a good tradeoff?

A sensible way to tune the alarm is to make it just sensitive enough so that it causes an acceptable number of false alarms. For instance, we may decide that the alarm is triggered only one percent of the days when the car is not burglarized.

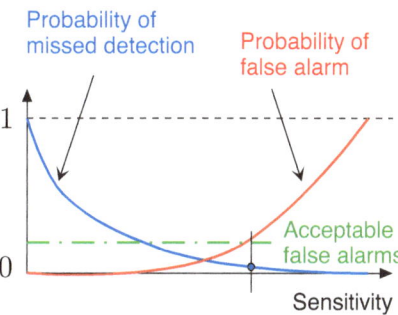

A similar method is used to design medical tests that achieve an acceptable tradeoff between false positives and false negatives.

Refinement

If you live in a very peaceful area where car burglaries almost never occur, you may want to reduce the sensitivity of your car alarm. The rational approach is to consider that a false alarm has a cost and so does a missed detection. You then want to minimize the cost that you accumulate over a large number of days. If the likelihood of a burglary is very small, the dominant cost becomes that of false alarms and you can reduce it by making the alarm less sensitive.

Guess my Weight

In many applications one needs an estimate of some unobserved quantity. For instance, the scheduler in a hospital needs estimates of the duration of procedures. The Human Resources department at Walmart™ needs estimates of the workload in order to staff its stores. The Department of Transportation uses estimates of the traffic when planning road expansions.

To explain a very useful technique, let us consider the problem of estimating my weight. The only information that I give you is my height: 69 inches.

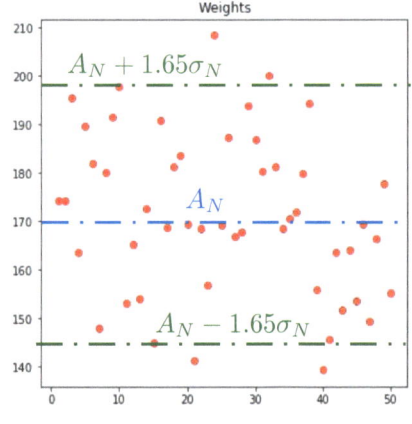

You have a list of weights $X_1, \ldots, X_N$ of N men who are 69-inch tall. These are indicated by the red dots in the figure. Here, $N = 50$.

You compute the average A_N of these weights. That is, $A_N = (X_1 + \cdots + X_N)/N$. This quantity is called the *sample mean*. Here, $A_N \approx 171$. Since N is quite large, you expect A_N to be a good estimate of my weight. However, as we can see in the figure, the weights tend to differ quite a bit from the average A_N. Accordingly, it is quite likely that my weight differs from A_N as well.

We can measure the importance of these differences as follows. First, we compute

$$\sigma_N^2 = \frac{1}{N}[(X_1 - A_N)^2 + \cdots + (X_N - A_N)^2].$$

Thus, σ_N^2 is the mean squared deviation between the weights $X_1, \ldots, X_N$ and their average A_N. One calls this quantity the *sample variance* of the weight. The square root σ_N is called the *sample standard deviation*.

The meaning of σ_N is the *typical deviation* between a weight X_n and the average A_N. If all the weights had been very close together, σ_N would have been small. In our example, $\sigma_N \approx 16$. This suggests that it is likely that my weight differs from the average 171 by some small multiple of 16.

The theory says that, with N reasonably large (larger than 20), the likelihood that my weight is in the interval

$$[A_N - 1.65\sigma_N, A_N + 1.65\sigma_N]$$

is about 90%. Thus, this interval is a 90%-confidence interval for my weight. Here, this interval is $[145, 197]$. This interval is very wide, but this is the best you can do. Fortunately for me, my weight is indeed in this range.

In practice, it is essential to be aware of the confidence interval and not assume that the average A_N is a precise estimate.

Note that choosing a larger value of N does not improve your estimate significantly. It is important for N to be reasonably large, but it does not help to use a very large value of N. The reason for this is that there is some irreducible randomness in the weights, and looking at many weights will not improve significantly your guess of my weight.

 ## How does the weight depend on the height?

One is often interested in the factors that affect some quantity. For instance, does the number of books a teenager reads affect strongly his GPA as a senior in college? Does the number of hours he spends playing video games? How about the number of hours he sleeps? Also, is the SAT a good predictor of the college GPA?

To explore this question, let us try to understand how someone's height affects his weight. We start by looking at a set of (height, weight) pairs for a number of men. For each man, the pair is shown as a red dot in the figure.

As the figure shows, and as we expected, taller people tend to be heavier. To get a sense of the dependency of the weight on the height, we draw a straight blue line through the middle of the red dots, trying to be as close as we can to the dots. The line goes up, showing that the weight tends to increase with the height. The slope of the line indicates the strength of the dependency. If the line had been almost horizontal, this would have indicated that the weight does not depend significantly on the height. Here, the slope is about 3.5 pounds per inch. Thus, a man who is 10 inches taller is 35 pounds heavier, typically. This does not mean that to lose weight you should try to get shorter!

This blue line is called the *linear regression* of the weight over the height. There is a systematic method for drawing the line, and we explain it below. For now, let us explore a more important question: how confident are we in this line? Could

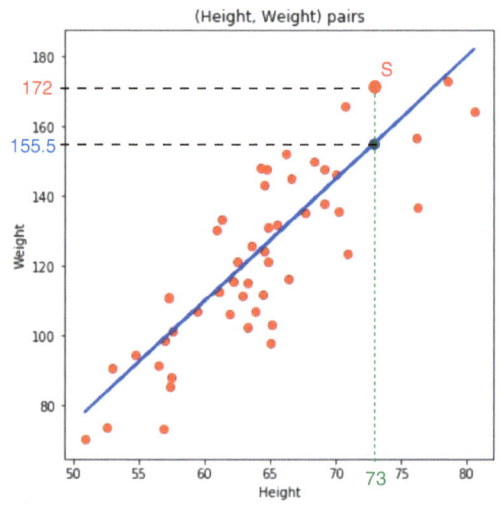

it be that if we look at another set of men we get a very different line? In other words, is the 3.5 pound per inch universally representative, or is it specific to the samples that we looked at?

As you suspect, the quality of this line depends on the number of sample points and on the degree of variability of the weights as a function of the height. For instance, if you look at only two sample points, then the line that you draw through these two points is not representative, as you can see by picking two red dots at random in the figure. In practice, how can you tell if you have enough samples? Here is a pragmatic approach. Say that you have N points. Choose a subset of half of the points at random. Draw the blue line. Then add a new point chosen at random, draw the new line and continue in this way. If the successive blue lines that you obtain become more and more similar, then you can trust that you have enough samples. Otherwise, you probably don't.

How does one draw the blue line? This may not matter much, as standard programs such as Excel, Python, Matlab, and R do it for us. However, the discussion clarifies the precise meaning of the linear regression.

The idea is to minimize the sum of the squares of the vertical distances between the red dots and the blue line. To see what this means, look at point S in the figure. It corresponds to a man with height 73 inches and weight 172 pounds. The blue line gives a weight equal to 155.5 pounds for a height of 73 inches. The square of the error is $(172 - 155.5)^2$. The blue line is the one that minimizes the sum of these squared errors over all the samples.

Mathematically, the blue line is a relation $W = a + bH$ that computes an estimated weight W for a person with height H. The parameters a and b determine the line. In this example, $a = -100$ and $b = 3.5$. Thus, for a person with height 73, the blue line gives an estimated weight $W = -100 + 3.5 \times 73 = 155.5$. If H_n is the height of the n-th man in the set of samples and W_n is his weight, the blue line gives an estimate $a + bH_n$ instead of W_n. The squared error for the n-th sample point is then $(W_n - a - bH_n)^2$. The parameters (a, b) that determine the blue line are those that minimize the sum of these squared errors, i.e.,

$$(W_1 - a - bH_1)^2 + \cdots + (W_N - a - bH_N)^2.$$

Some algebra enables to compute a and b.

5
Gambling

Is it foolish to play the lottery? Winners say No.

Gambling is an interesting way to test our relationship to un-certainty. We start with a toy example and then explore more realistic situations.

 ## A Toy Example

Your friend flips a 'fair' coin that is equally likely to produce 'heads' or 'tails.' You will get $10.00 if the outcome is 'heads' and nothing if it is 'tails.' How much are you willing to pay to play this game?

A sensible answer seems to be 'any amount less than $5.00.' One justification is that if you play the same game 100 times, you expect to win about 50 times, or a total of $500.00. Thus, your net profit should be positive if you pay less that $5.00 every time you play the game. In other words, you expect to win $5.00 on average when you play the game, so you should be willing to pay any amount less than $5.00 to play.

 ## Does this Make Sense?

Let's explore this answer a bit further. Say that you can play the game only once, not 100 times. Does that change your reasoning? Now, assume that the reward is $100,000.00 instead of $10.00. Would you be willing to pay $49,000.00  to play the game? Would you bet your house if you would lose it or win a second one with even odds?

 ## Utility?

.

I would not bet my house. What is going on here is that the loss in 'utility' of losing my house is much larger to me than the gain in utility of winning a second house. Thus, my expected gain is very negative and I don't want to play the game.

For a similar reason, I may buy a lottery ticket. My expected gain in dollars is negative, but it is positive in utility. Indeed, losing the cost of a ticket has a negligible impact on my welfare, but winning the jackpot would probably have a big positive effect. This is why we insure our houses even though this results in a expected financial loss (otherwise, insurance companies would not exist).

The same logic explains that you should not buy an extended warranty for a $200.00 printer, but that you might want to do it for an expensive car.

Thus, a reasonable model of behavior is that we try to maximize our expected *utility*. We recognize that the utility is not simply proportional to our 'fortune' in dollars, but it is a function of that fortune and we try to maximize the expected value of that utility. This is the model that many economists accept; it explains many of our 'rational' decisions.

Some authors argue that 'you should never buy a lottery ticket'. Their point is that you lose money, on average, every time you do. The same argument would lead you to never buy any insurance or warranty. This argument is too naive because it ignores the non-linearity of utility. By the same token, you should never get a cappuccino because the expected financial gain is certainly negative. It is quite possible that money is not our only goal in life.

Even though the utility maximization model is useful, we are more complex than simple 'utility maximizers'. Indeed, many experiments demonstrate that we do not behave in a way that is consistent with utility maximization, as we discuss next.

## Choosing a Car

Say that you are in the market for a new car and that you are considering a Corolla, a Camry, or a Lexus. You prefer the Camry to the Corolla, because it is not much more expensive and yet offers more comfort. For the same reason, you prefer the Lexus to the Camry.

However, when you compare the Corolla and the Lexus, you prefer the Corolla, because it is substantially less expensive than the Lexus and you cannot justify spending that much more for a car.

Although this situation may not apply to you, you can probably easily imagine a similar one that does. Such examples show that our preferences are not always transitive. That is, we may prefer A to B and B to C, but C to A. Consequently, we cannot assign a numerical value to the 'utility' we derive from each option, because if we could, the preferences would be transitive.

Thus, if you compare the three cars two by two, your choices seem inconsistent. What do you do when you, in fact, compare the three cars together?

A similar situation arises when you have picked your car and the dealer suggests options. Do you want the navigation system? It is only $400.00 and so convenient. Sure. How about the 7-speaker audio package. It is only $350.00 and sounds so much better. Sure. What about leather seats? Sure. And I recommend the Xenon lights. Sure. Pretty soon, you have selected options that add $3,200.00 to the cost of the car. If you had only two choices: standard and loaded, you might have gone for the standard model.

 ## How to Choose?

These examples show that there is no simple way to formulate our choices. Even if we believe that we are maximizing our utility, we never really compute it. Instead, we act intuitively and probably in a very inconsistent way.

Nevertheless, it is possible to be more rational than we are in our choices. A sensible first step is to be aware of the risks. A second step is to think of your perceived utility. Of course, it is perfectly fine to be inconsistent and somewhat irrational. However, it may be dangerous to be oblivious when making choices.

 ## Lottery

Games of chance motivated the early work on probability and continue to inspire research. Some games are particularly deceiving and are designed for us to have a vastly incorrect appreciation of the odds of winning.

Consider a lottery game where you have to select 6 numbers out of 70. If you guess correctly, you are rich. Now, 6 out of 70 sounds about like 1 out of 10, so winning should not be too unlikely.

Unfortunately, the odds of winning are much smaller than 1 out of 10. It turns out that they are about one in 131 million. To verify this fact, note that when you pick numbers, you have 6 chances out of 70 that your first choice is a winning number. Also, if your first choice is a winning number, you now have 5 chances out of 69 that your second choice is again a winning number. This is similar for the other numbers. So your chances of winning are $(6/70) \times (5/69) \times (4/68) \times (3/67) \times (2/66) \times (1/65) \approx 1/(131\text{million})$.

What makes it so unlikely is that you have to be quite lucky 6 times in a row. Even though the odds that each number is a winning number are between $6/70$ and $1/65$, the odds of being lucky 6 times in a row when picking the 6 numbers are very small.

If you were to buy all the possible lottery tickets, for $1.00 each, you would spend $131M and would be sure of winning. However, one or more other players might also win. Your actual winnings could then be the jackpot or only a fraction of it. Note that if the jackpot is less than $131M, you are certain to lose money, on average, when you buy a ticket. However, as we discussed earlier, it may still be rational for you to play, depending on your utility.

Slot Machine

Another deceiving game is the slot machine. A highly simplified model of this game is that you insert $1.00 in the machine and that you lose your bet with probability 0.51 and otherwise get $2.00 back, thus gaining $1.00 with probability 0.49.

Say that you play this game 1000 times. You expect to lose $1.00 about 510 times and to gain $1.00 about 490 times. Thus, you expect to lose $20.00 after playing 1000 times. Not too big a deal. You can have quite a bit of fun for $20.00. Also, it seems that losing $20.00 after gambling 1000 times is a small probability of loss. After all, you might hit a lucky streak and win the majority of those times, so that you could come out of the casino with hundreds of dollars more than when you went in.

A closer look at the game reveals that you may not be frequently lucky. Say that you enter the casino with $100.00 in your pocket and that you will play the game until you ei-

ther lose everything or until you accumulate $500.00. What are the odds that you come out with $500.00? At first glance, you might guess about 1 in 5. In fact, these would be the correct odds if the slot machine were fair, i.e., if it let you win 50% of the time. However, since you win only 49% of the time, it turns out that the odds of you getting out with $500.00 are only about 1 in 9 million. Thus, if you were to go to the casino every day and re-peat this process every time, you might have to do it for about 25 thousand years before you have a winning day.

Why is this game so deceiving? The reason is similar to the lottery. You have to be a little bit lucky many times for your winnings to reach $500.00. You are struggling against the odds. Thus, although the slot machine takes only $20.00 out of every $1000.00 that are played, there are very few rich slot machine players.

If you could play $100.00 at a time instead of $1.00, then the odds that you would get out with $500.00 would be about 18%. This example shows the role of 'table limits' in casino games. Also, if you were more reasonable in your expectations and could stop after reaching $120.00 instead of hoping for $500.00, you would be much better off.

 ## Investments: Kelly Strategy

A relative gives you a sum of money. You decide to invest it and try to make it grow. Say that there are two financial instruments you can use: A and B. Instrument A is a treasury bond that guarantees a fixed return of 5% per year. Instrument B is a very risky stock that either goes bust with probability 90% or gets multiplied by 20 with probability 10%, in any given year. How should you invest? I explain the Kelly strategy that is supposedly used by some of the most successful investors.

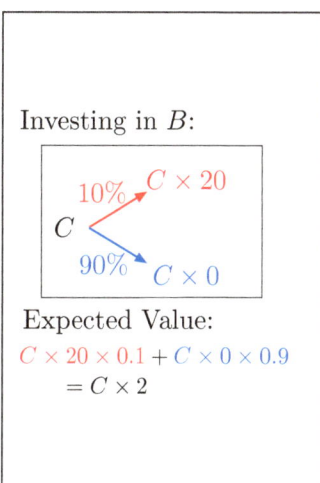

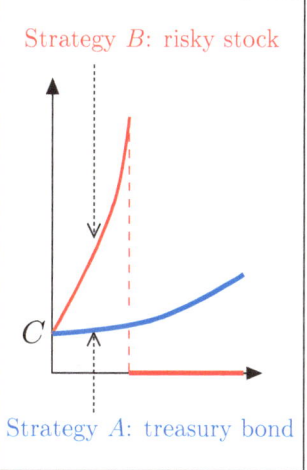

Say that you start with a capital C. If you invest in A, it grows to C × 1.05 after one year. If you invest in B, it becomes 0 after one year with probability 0.9, and 20C with probability 0.1. Hence, if you invest in B, the expected value of your capital after one year is $0 \times 0.9 + 20C \times 0.1 = C \times 2$.

Thus, if your goal is to maximize the expected value of your capital after one year, you should invest in B. The same argument holds for every subsequent year. Thus, to maximize the expected value of your capital after any given number of years, you should invest in the risky stock.

Your wise uncle tells you that this strategy is foolish. Indeed, you are almost certain to lose all your money in a few years. The right-hand side of the figure above plots your capital over the years with the two strategies and shows that, inevitably, your capital when investing in A eventually dominates what you would have by investing in B.

However, by investing in A, you must be missing on the opportunity for the fast growth of investment B. Is there a way to capture some of that growth while still playing it relatively safe?

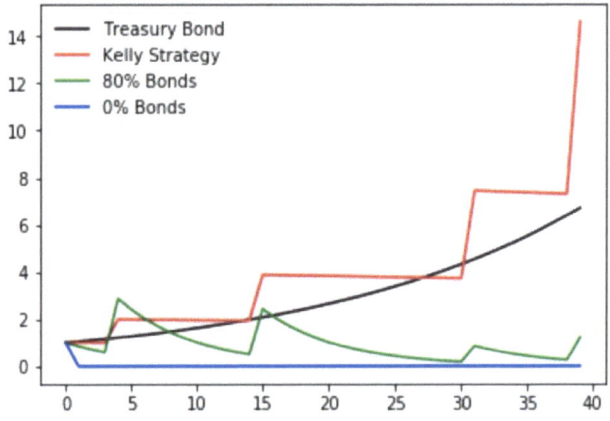

John Kelly proposes an investment strategy that combines the safety of investment A with the potential high growth of investment B. He suggests to invest a fraction f in A and the remaining fraction $1 - f$ in B. Also, he explains how to calculate the best fraction f you should invest in treasury bonds. In this example, his calculation shows that $f = 95\%$, so that you should invest 95% in treasury bonds and 5% in the risky stock. The figure above shows typical evolutions of your capital for different strategies. To find the best fraction f, one maximizes the *doubling rate* of the fortune, defined as the logarithm of the expected multiplier.

A word of caution: this strategy makes sense if you repeat it many times. If you use it only a few times, there is a significant chance that investing in *A* may outperform the Kelly strategy. Kelly's strategy has been used effectively in the stock market where investment decisions are frequent.

6

Randomized Algorithms

> *I have a giant baking book, so I close my eyes and pick a random page. Whatever it is, I try to bake it!*
>
> Nina Dobrev

There are many situations where we make random choices, on purpose. A coin flip determines who serves first in a tennis game. One draws at random the competitors for the world cup of soccer. The goal of these random choices is to guarantee fairness. My sister has a robotic lawnmower that moves in a random direction after it hits an obstacle. The robot ends up mowing the lawn remarkably evenly. Alexa™ selects her daily jokes randomly, which makes it unlikely that the same joke repeats within a few months. I tend to look for my car keys at random, probably increasingly.

Intuitively, randomized algorithms can be useful, but is that really the case? We first explore how a computer makes random choices. Then we examine a few randomized algorithms. We conclude the chapter by exploring whether randomness is fundamentally useful.

 Generating Randomness

It is somewhat puzzling that a deterministic computer can make random choices. One could equip the computer with a Geiger counter or a device that measures thermal noise to have access to truly random quantities. This is not what is done.

Instead, one designs a *random number generator* that works as follows. One chooses a function $f(\cdot)$ and, starting with some value x_0 called the *seed* of the generator, one computes successively $x_1 = f(x_0)$, $x_2 = f(x_1)$, $x_3 = f(x_2)$, etc. Many functions $f(\cdot)$ have been proposed and the sequences $\{x_0, x_1, \ldots\}$ that they produce have been tested for 'randomness'. A function that is widely used is the 'xorshift' function. It computes the bit-wise x-or of x and its left-shifted version. The left shift amplifies the impact of a small change in x_0 on subsequent values, thus creating a chaotic dynamic.

How does one test a sequence of numbers for randomness? Consider a sequence of 0's and 1's, say 0100110.... What does it mean to say that it is random? The idea is to check if the sequence could have produced by coin flips. For instance, one may want to see if the fraction of 1's is close to 50%, if the fraction of times that 0010 appears is the same as in a sequence of coin flips, and so on. If the answers to those questions are affirmative, we agree that the sequence looks random.

The quality of a random number generator matters substantially when we use that generator to simulate systems with random inputs or randomly generated choices. Considerable research has been done on the testing of random number generators. It is not easy to fake coin flips, but today's best generators come very close.

Is this good enough? Imagine a complex system, say a nuclear power plant, that happens to fail when the random input is a specific sequence 0010010111001. If the bits correspond to coin flips, this specific pattern may occur once every three years, which makes the design of the power plant unacceptably dangerous. What if the random number generator used to simulate the system has the peculiar defect that this sequence only occurs once every millennium? The simulation would erroneously conclude that the failure of the plant is so rare that it is acceptable. The point of this contrived example is that the validity of a random number generator depends on what one does with it. If one knew the critical sequences, one could test for them, but the point of the simulation may be that one does not know these sequences.

Estimating an area

Imagine a complex figure A defined in a square with unit sides by some functions and one would like to estimate the area of A. The geometry of the problem is complex and the calculation is not easy. Consider the following randomized algorithm. One chooses points randomly, uniformly in the unit square and one counts the fraction of points that land inside A. This fraction converges to the area of

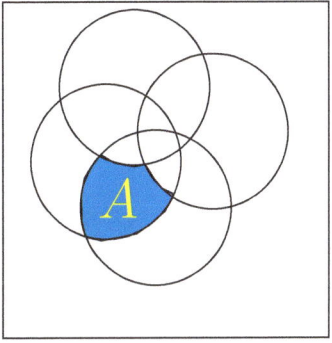

A. Moreover, as we saw in the chapter *Causation*, with one thousand points, the fraction is within 3% of the area, with 95% confidence. This procedure is called a *Monte Carlo simulation*.

Figuring out if a point is inside A is straightforward. In the example of the figure, one simply checks the distances from that point to the centers of the four circles to determine which circles the point falls in. Thus, this algorithm takes only a few instructions. Moreover, one has some guarantees about its accuracy. The same idea works to estimate a volume in any number of dimensions.

A deterministic algorithm could use a grid of points in the square, say a regular grid with 60×60 points. Imagine that the area A is a rectangle. The grid would determine the sides of the rectangle with a worst-case error of $1/60 \approx 1.5\%$. Since the two sides are known $\pm 1.5\%$, the area is computed $\pm 3\%$, which is the accuracy of the randomized algorithm (with 95% confidence). This deterministic algorithm requires $60 \times 60 = 3600$ points, which is about the same as the 1000 points of the randomized algorithm.

However, consider the same problem in d dimensions. The accuracy of each side should be $(1/d) \times 3\%$ for the volume to be calculated $\pm 3\%$. This means that each side should have $d/(3\%) \approx 30d$ points, for a total of $(30d)^d$ points. The complexity of this deterministic algorithm grows exponentially fast in the dimension of the object, whereas that of the randomized algorithm remains fixed.

Random Search

Assume we have $2n$ boxes for some $n \geq 1$, half of which contain a red marble, and the other half contain a blue marble. We want to find a red marble.

A simple deterministic approach is a sequential search: look at box 1, then box 2, and so on until you find a red marble. In the worst case, if the first half of the marbles are blue, this algorithm takes $n + 1$ steps.

A simple randomized algorithm is as follows. Choose a box at random. If it contains a red

1 2   2n

marble, stop; otherwise, repeat. For $m = 1, 2, \ldots$, the probability that we need m steps is $(1/2)^m$. Indeed, the proportion of blue marbles is $1/2$ at each step, so that the probability of finding $m - 1$ consecutive blue marbles and then one red marble is $(1/2) \times (1/2) \times \cdots \times (1/2) = (1/2)^m$. The expected number of steps is then equal to

$$1 \times (1/2) + 2 \times (1/2)^2 + 3 \times (1/2)^3 + \cdots = 2.$$

Moreover, the probability that it takes more than m steps is $(1/2)^m$ (the probability that the first m marbles are blue), so we know that the algorithm is likely to terminate quickly.

 ## Are randomized algorithms useful?

Is our intuition correct that randomization is useful in some algorithms? Can this intuition be verified precisely? Let's explore this fact with a bit more care in the search example.

Any deterministic sequence of boxes that we explore takes $n + 1$ steps, in the worst case where the first n boxes we open all contain blue marbles. In contrast, the randomized algorithm has probability $2^{-10} \approx 0.1\%$ of taking more than 10 steps, independently of the number $2n$ of boxes. In that sense, the randomized algorithm is 'safer' than any deterministic algorithm.

However, note that for any given sequence of opening the boxes, the worst case arrangement of marbles is when the blue marbles are in the first n boxes one opens. Thus, one could argue that the worst case for the randomized algorithm also takes $n + 1$ steps. It seems to me that the distinction between the deterministic and the randomized algorithms is quite subtle. If the statement is 'give me your deterministic or your randomized algorithm and I then will arrange the marbles,' then, trivially, the randomized algorithm is better. However, if one compares the worst case of both algorithms over all the possible arrangements of marbles, then they are equivalent.

You could imagine a spy checking your code, then arranging the marbles. If you choose the randomized algorithm and the spy cannot see the seed x_0 of your random generator, the randomized algorithm is safer. If he can see x_0, then he sees a deterministic algorithm.

The point of this discussion is that one has to be careful about the meaning of 'worst case' and 'randomization'.

7
Auctions

> *Publication - is the auction of the mind...*
>
> Emily Dickinson

I find auctions fascinating because they ask you to think about how other people think. Whether you win an auction depends on how much other people bid. When you choose your bid, you try to guess how much other people will bid. They do the same and they know that you are thinking about how they think.

To explain some important ideas about auctions, I discuss the second price and first price auctions.

 ## Second Price Auction

You are in an artworks auction. You want to bid on an etching by Georges Braque. The mechanism is an *ascending auction* and it proceeds as follows. The auctioneer opens the bidding at $800.00. Ten people in the room, including you, raise their hand to indicate they are willing to pay at least $800.00 for the artwork. The auctioneer then raises the price by $100.00. The process continues until there is only one bidder, Fred, who raises his hand when the auctioneer's price reaches $3,200.00. That lonely remaining bidder, Fred, then gets the etching for $3,200.00.

Now, consider the following modification of the auction, called a *second price auction*. Every bidder is asked to put a bid in a sealed envelope and give it to the auctioneer.

The rule of the auction is that the highest bidder gets the item, but pays only the second highest bid. In the ascending auction, we know that Fred values the etching more than $3,200.00, say $3,800.00. We also know that there was at least another bidder, say Sam, who valued the etching more than $3,100.00 but less than $3,200.00, say $3,150.00. Thus, if all the bidders write in their true valuation, Fred will be the highest bidder and bid $3,800 and Sam will be the second highest bidder and bid

$3,150.00. Thus, Fred gets the item and pays $3,150.00. This is essentially the same as the ascending auction, up to the quantization of the steps.

In the ascending auction, it is obvious that the bidders should be *truthful*. That is, they should not stop bidding while the auctioneer's price is still less than their valuation, for they would then miss out on an item they could get for less than their valuation. Also, they should not keep bidding when the auctioneer's price exceeds their valuation, for they would then pay more for the item than their valuation.

Since the second price auction is essentially identical to the ascending auction, the bidders should also be truthful and submit their true valuation.

 ## First Price Auction

You want to buy a house in Berkeley, California. The rule is that you write an offer that will compete with a number of other offers. The bidder with the highest offer gets the house and pays the amount of the bid. To be acceptable, your offer should exceed a

value called the *reserve price* or *asking price*.

The agent of the seller tells you how many other people are bidding on the same house. (In practice, all the potentially interested parties find out the number of bidders before they place their actual bid.) How much should you bid?

Say that you value the house at $1M and that the asking price is $900k. If you bid $1M and are the highest bidder, you get the house and pay exactly how much you value it. In this sense, you make no profit. So, you try to *shade* your bid and offer less that $1M. For instance, say that you offer $940k. You hope that no other bidder will offer an amount between $940k and $1M, because you would then miss out on the house even though you might have had it for less than $1M. So, you want to shade your bid just enough.

The goal is to maximize your *expected profit*, defined as the profit multiplied by the probability that you get the house. Here, the profit is defined as the difference between your valuation and the amount you pay. For instance, say that, if you bid $940k, the probability that you get the house is 10%. Since you value the house at $1M, your expected profit is ($1M − $940k) × 10% = $6k. Assume also that, if you bid $960k, the probability that you get the house is 50%. In this case, your expected profit is ($1M − $960k) × 50% = $20k. In that sense, bidding $960k is preferable to bidding $940k because the expected profit is larger.

Assume that there are N other bidders, in addition to you. Assume also that the valuations of the N other bidders are uniformly distributed, say between $900k and $1.3M and your valuation happens to be $1M. Analysis show that if $N = 1$ you should bid $950k. If $N = 2$, you should bid $967k. In general, you should shade your bid by $100k/(N+1)$.

Revenue Equivalence

Which auction is best for the auctioneer? Is it the first price or the second price auction? Remarkably, they yield the same expected revenue for the auctioneer.

In the second price auction, the amount that the auctioneer gets is the second highest valuation. In our example, the auctioneer gets \$3,200.00 for the Braque etching even though Fred values it at \$3,800.00. In the first price auction, one can also show that the auctioneer gets the same as in a second price auction. This happens because the bidder with the highest valuation shades his bid. The expected value of the shaded highest valuation turns out to be exactly equal to the expected value of the second highest valuation.

Let's verify it in the case of two bidders who have a valuation that is uniform in $[a, b]$.

The figure shows the interval $[a, b]$ and two valuations V, W picked at random uniformly in that interval. The circle is another way to look at this selection. Pick three points: black, red, blue, uniformly at random on the circumference of the circle. Call the black point (b, a), the red point V, and the blue point W. Now, break the circle at (b, a) and unfold it: you end up with the top figure with the interval. The symmetry of the choices on the circle shows that

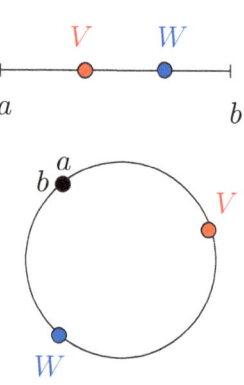

$[a, V], [V, W]$, and $[W, b]$ have the same expected length, by symmetry. Moreover, this common expected length must then be equal to $(b - a)/3$. Hence, the expected value v of the second highest valuation is such that $(v - a) = (b - a)/3$, i.e., $v = a + (b - a)/3$.

In the first price auction, the bidder with the highest valuation W will shade his bid and offer only $a + (W - a)/2$, as we saw in the discussion of this auction (here, $N = 1$). Now, the figure also shows that the expected value w of the highest valuation W is such that $(w - a) = 2(b - a)/3$. Thus, the expected highest bid is $x = a + (w - a)/2 = a + (b - a)/3$, which is the same as v, the expected bid for a second price auction.

You can verify that the same result holds for $N + 1$ bidders.

 ## Reserve Price

The first and second price auctions yield the same expected revenue for the auctioneer, but this revenue can be improved. To see this, consider a first price auction with a single bidder with a valuation picked uniformly in $[0, 1]$. Since the bidder is the only one, he can bid 0 and get the item for free.

Now assume that the auctioneer sets a minimum bid – a reserve price – equal to r. The single bidder bids his valuation v and gets the item for r if $v \geq r$; if $v < r$, the bidder does not get the item. In this case, the expected revenue of the auction is $r(1 - r)$ since $1 - r$ is the probability that $v \geq r$. This expected revenue is maximized by choosing $r = 0.5$. The reserve price that maximizes the revenue of the auction can be calculated in the general case, but the analysis requires some more complex algebra.

8
Games

> *We do not stop playing because we grow old, we grow old because we stop playing!*
>
> Benjamin Franklin

In many games, the optimal way to play is to choose the action at random. For instance, you play soccer and are about to shoot a penalty. To try to fool the goalie, you choose the target of the ball in an unpredictable way and avoid any regular pattern. The same is true when you serve a tennis ball, and in many other situations.

Thus, games are simple examples where uncertainty improves the strategy. I find this fact interesting because we normally associate rationality with identifying the best course of action and not with making random choices.

I discuss a few familiar games to illustrate some key concepts.

Matching Pennies

Alice and Bob play this game by both simultaneously putting down a penny on the table. If the faces match, Bob gives Alice $1.00; otherwise, Alice gives Bob $1.00. This is a simplified version of the penalty shot where Bob shoots and Alice defends. If the direction that Alice chooses matches that of Bob, then Alice wins; otherwise, Bob does.

Say that Alice tends to favor 'heads', i.e., chooses 'heads' with a probability larger than 0.5. Then, Bob can always choose 'tails' and he wins more than half of the time. In fact, choosing 'tails' is then the decision that maximizes Bob's odds of winning. Similarly, if Alice favors 'tails', then Bob chooses 'heads' and wins most of the time. Thus, to minimize her losses, Alice should choose 'heads' with probability 0.5.

Thus, the best strategies are for Bob and Alice to choose both faces with equal probabilities. The best choices are indeed random. This pair of strategies for Alice and Bob is called a *Nash equilibrium* (after John Nash), which means that no player has an incentive to deviate unilaterally from it. Nothing too exciting here. But we will see that simple variations quickly become more interesting.

 Alice's Serve

Alice serves at tennis and Bob receives.

For simplicity, say that Alice has two options: wide or middle. However, Alice's wide serve is much better than her middle serve. If she serves wide, and Bob guesses correctly, Alice wins the point 65% of the time. However, if Alice serves in the middle and Bob guesses correctly, Alice only wins the point 55% of the time. In both cases, if Bob guesses incorrectly, Alice wins the point.

How should Alice and Bob choose in order to maximize their

odds of winning the point? Intuitively, Alice should favor her wide serve since it is her stronger shot. However, Bob knows that. Should she then go for a weaker shot? But then, Bob would know that she will favor that shot, ...

To study this game, let p be the probability that Alice chooses the wide serve. Bob then argues as follows. If he guesses wide, Alice wins the point with probability 65% when she serves wide, and with probability 100% if she serves in the middle. Thus, if Bob guesses wide, Alice wins with probability $p \times 0.65 + (1 - p) = 1 - 0.35p$. Similarly, one finds that if Bob guesses middle, Alice wins with probability $p + (1 - p)0.55 = 0.55 + 0.45p$. Bob then selects the guess that minimizes the probability that Alice wins the point. The result is that the probability that Alice wins the point is the minimum of $1 - 0.35p$ and $0.55 + 0.45p$. To maximize this minimum, Alice should choose the value of p such that $1 - 0.35p = 0.55 + 0.45p$, i.e., $p \approx 0.56$. With this selection, Alice wins the point with probability $1 - 0.35 \times 0.56 \approx 0.8$, whether Bob guesses wide or middle.

Does the choice of Bob matter? Intuitively, if Bob tends to guess wide, Alice should favor serving middle. To understand this more precisely, assume that Bob guesses wide with probability q. If Alice serves wide, she wins the point with probability $q \times 0.65 + (1 - q) = 1 - 0.35q$. If she serves middle, she wins the point with probability $q + (1 - q) \times 0.55 = 0.55 + 0.45q$. Thus, she should make the choice that maximizes that probability and she then wins

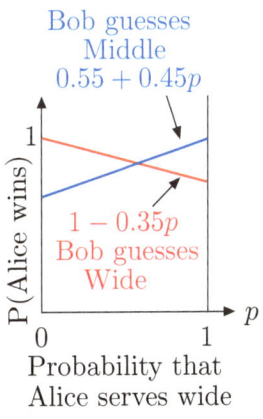

Bob guesses Middle $0.55 + 0.45p$

$1 - 0.35p$ Bob guesses Wide

P(Alice wins)

Probability that Alice serves wide

the point with a probability equal to the maximum of $1 - 0.35q$ and $0.55 + 0.45q$. Bob should then choose the value of q that minimizes this maximum. It is the value of q such that $1 - 0.35q = 0.55 + 0.45q$. Thus, $q \approx 0.56$. For this choice of q, Alice wins the point with probability 0.8, no matter how she serves.

Here, the choices $p = 0.56$ and $q = 0.56$ are a Nash equilibrium. Moreover, any other pair of values (p, q) is not an equilibrium. Indeed, if $p < 0.56$, then the best choice for Bob is $q = 1$ and if $p > 0.56$, the best choice for Bob is $q = 0$. Similarly, if $q < 0.56$, the best choice for Alice is $p = 1$ and if $q > 0.56$, the best choice for Alice is $p = 0$.

The conclusion of our analysis is that Alice should serve wide with probability 56% and that Bob should guess wide with probability 56%. Alice then wins the point with probability 80%. So, Alice should indeed favor her best serve, but not excessively. Similarly, Bob should guess that Alice will favor her best serve, but not always. These results confirm our basic intuition, and make it precise.

Reinforcement Learning

In practice, Bob may not know the strongest serve of his opponent. Bob can learn by observing the success rate of Alice when she serves wide and he guesses correctly, and similarly when she serves middle. Eventually, Bob can figure out the probabilities 65% and 55%, and then compute the optimal probability 56% of guessing wide.

This method seems rather convoluted and is not likely to be the method that tennis players use. Instead, one can imagine a simpler idea called *reinforcement learning*. This method consists in doing more of what works well and less of what does not work well.

Let us explore such a scheme where Alice increases the probability of serving wide when she is successful when doing so or when she fails when serving middle. Similarly, Alice decreases the probability of serving wide when she is not successful when doing so or when she succeeds when serving wide. Bob uses a similar strategy. By increasing p, we mean replacing its value by $0.9p + 0.1$ and by decreasing p we mean replacing its value by

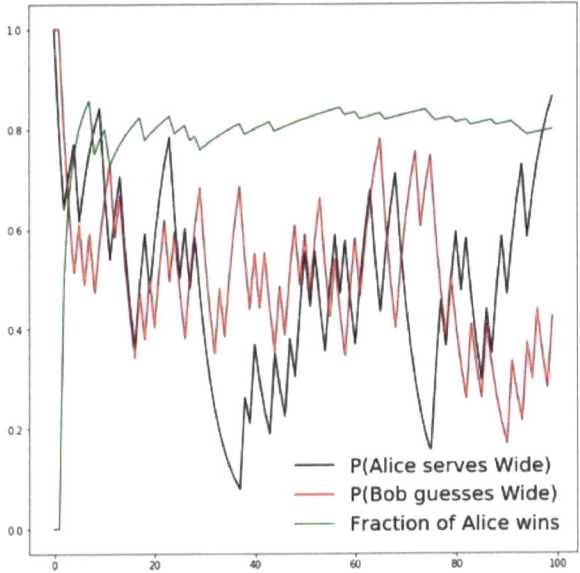

0.9p, and similarly for q.

The figure shows that when using this procedure, the long term fraction of time that Alice wins converges to the optimal value 0.8 of the game, even though the probabilities of serving and guessing wide oscillate wildly.

9
Planning

We often face a delicate choice between short-term and long-term rewards. We give up smoking, work out at the gym, and frequently skip a second serving or a dessert hoping to prolong our healthy life. We go to school to improve our future prospects. The choice is delicate because of the uncertainty of the outcomes. We explore a systematic way to think about such choices.

Choosing games

The figure illustrates a problem of *sequential decisions*.

The main feature of this problem is that choices you make affect your future options. The figure shows a situation where you have to choose games to play.

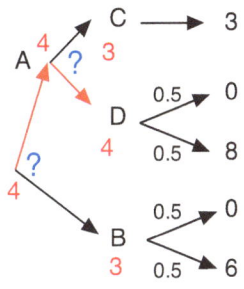

The first choice is indicated by the left-most question mark: you either choose game A or game B. If you choose game A, you face another choice: game C or game D. When playing game C, you win $3.00. When playing game D, you either lose with probability 0.5 or you win $8.00. When playing game B, you either lose with probability

0.5 or you win $6.00. How should you choose, assuming that you want to maximize your expected reward?

The solution procedure is as follows. The key idea is to start with the last choices, instead of with the first ones. That is, we proceed backwards instead of forwards. If you end up playing C, you win $3.00. Let us indicate next to C that the reward is 3. If you end up playing D, your expected reward is $0.5 \times 0 + 0.5 \times 8 = 4$. We write that expected reward next to D. Now assume you play A. You have the choice between C, which has reward 3, and D, which has expected reward 4. Thus, when you play A you should choose D and the expected reward of A is then that of D, i.e., 4. We write that expected reward next to A. Similarly, the expected reward of B is 3. Now consider the first choice: A or B. If you choose A, the expected reward is 4; if you choose B, the expected reward is 3. Hence, you should choose A. Also, the expected reward of the overall selections is 4.

The procedure that we just went through is called *stochastic dynamic programming*. We started with the last step and evaluated the expected reward we would get if we were to reach that step. We then continued backwards, step by step, until we reached the first step. At each step, we make the choice that maximizes the expected reward.

 ## Guess an ace

Consider a deck of 52 cards, 4 of which are aces. I turn over the cards, one at a time. At each step, before I turn over the next card, you may decide either to stop or to continue. If you stop and the next card is an ace, you win; if it is not, you lose. How should you play the game to maximize your chances of winning? Should you let a few cards go by and stop when the proportion of aces that remain in the deck is large enough?

The *state* of the deck is summarized by two numbers m and n where m is the number of aces and n is the number of cards that remain in the deck. If the state is (m, n), and you decide to stop, the next card is an ace and you win with probability m/n. If you decide to continue, with probability m/n, the next card is an ace and the next state of the deck is $(m - 1, n - 1)$; also, with probability $1 - m/n$, the next card is not an ace and the next state of the deck is $(m, n - 1)$.

Assume that, for some given value of n, someone tells you that the maximum probability of winning is $V(m, n - 1)$ when the state of the deck is $(m, n - 1)$, for $m = 0, \ldots, 4$. We call $V(m, n - 1)$ the *value* of state $(m, n - 1)$. You can then calculate $V(m, n)$ as follows. If you stop, the probability of winning is m/n. If you continue, with probability m/n you face a new game with state $(m - 1, n - 1)$ and the maximum probability of winning is then $V(m - 1, n - 1)$; with probability $1 - m/n$, you face a game with state $(m, n - 1)$ and the

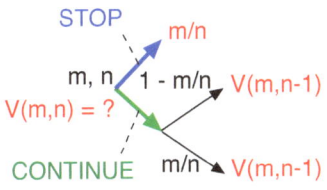

maximum probability of winning is $V(m, n - 1)$. Since the maximum probability of winning is the maximum over the two options (stop or continue), we find that

$$V(m, n) = \max\{m/n, (m/n)V(m - 1, n - 1) + (1 - m/n)V(m, n - 1)\}.$$

(In the expression, $\max\{x, y\}$ indicates the maximum of x and y.) You can verify that the solution of these identities is $V(m, n) = m/n$. Moreover, at each step, the expected reward is the same whether you stop or continue. For instance, we know that $V(0, 1) = 0$ and $V(1, 1) = 1$, so that

$$\begin{aligned} V(1, 2) &= \max\{1/2, (1/2)V(0, 1) + (1/2)V(1, 1)\} \\ &= \max\{1/2, (1/2) \times 0 + (1/2) \times 1\} = 1/2. \end{aligned}$$

Choosing a school

In the previous examples, the choices where free. In the next problem, we explore a model where choices have a cost.

Imagine that you can choose between two schools A and B. School A has a better reputation than school B, but it is more expensive. Is it worth it? Assume that a degree from A costs $200k whereas from B it costs only $100k. With a degree from A, the chances you get a job with Google™ are 30% and your expected lifetime earnings will be $2.5M, otherwise, your expected lifetime earnings will be $1M. With a degree from B, the chances of the Google™ job are 15%.

The figure shows that the expected earnings after going to school A are $0.3 \times 2.5 + 0.7 \times 1 = 1.45$ and they are 1.225 after going to school B. After subtracting the cost of the degree, we find that the best choice is school A.

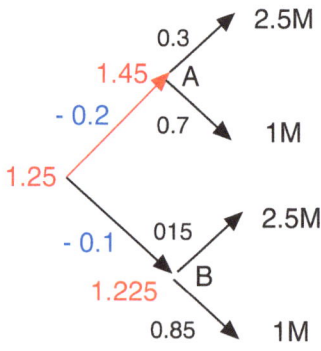

General Model

The previous examples are particular cases of a general model. In this model, one considers a system characterized by a *state* x. When one chooses an action a, the state of the system jumps to y. This new state y is chosen randomly with a probability $p(x, a, y)$ that depends on a and x, but not on the previous evolution of the system. Moreover, one incurs a cost $c(x, y)$ that depends on x and y. This process is repeated N times. The problem is to choose the action a at each of those N steps, based on the state x of the system at that step, to minimize the expected total cost over the N steps.

The number of problems that can be formulated in this way is quite amazing. This general method of stochastic dynamic programming has been used to solve problems as varied as optimizing the surgery schedule of a hospital and minimizing the fuel needed to land on Mars. As you can see, even the sky is not the limit.

 ## Principle of Least Action

In the previous examples, we explored how to minimize the sum of stepwise costs. Remarkably, the laws of physics correspond to the Principle of Least Action. This principle says that the motion of objects minimizes a sum of incremental costs called the *action*.

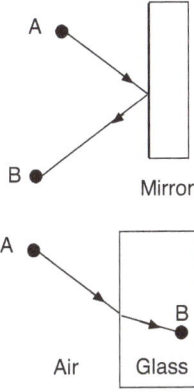

For instance, light travels along a path that minimizes the travel time between two points. In this case, the action is the travel time.

As another example, the Earth travels along a path around the Sun that minimizes the action.

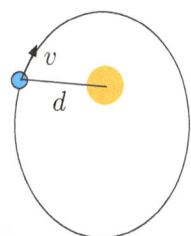

Here, the action is equal to the integral of the difference between the kinetic energy and the potential energy, i.e., the integral of $(1/2)mv^2 - P(d)$ where $P(d)$ is the potential energy of the Earth when it is at distance d from the Sun.

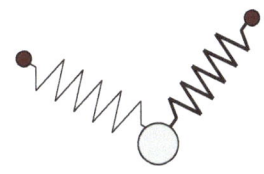

As a particular case, a system comes to rest in a state that minimizes its potential energy.

Around 1750, Pierre-Louis Maupertuis said that "nature is thrifty in all its actions" to encapsulate these observations. The Principle of Least Action applies to Quantum Mechanics and also to General Relativity.

10

Machine Learning

> *Fear artificial stupidity, not artificial intelligence.*
>
> Mark Bishop

Machine learning was a hot topic in 2019. The power of computers and new algorithms combine into an effective technology. This technology is demonstrated in systems as different as a computer beating Go masters, self-driving cars that are arguably safer than human-driven cars, and image and voice recognition systems that are remarkably effective (Alexa, play Debussy).

My goal in this chapter is to explain some of the key ideas of machine learning.

 ## Understanding, Intelligence, and Learning

Terminology can be confusing. One may think that intelligence refers to understanding. Bill is intelligent because he quickly understands difficult concepts. In that sense, learning means developing tools for understanding, and understanding refers to reducing a complex situation to a simple model that one can apprehend.

This is not at all what machine learning and artificial intelligence are about. In that context, intelligence means the ability to choose appropriate actions in novel situations.

In that sense, a thermostat is intelligent: it chooses the appropriate action (heating or cooling) based on the observed situation (the temperature in the room).

A learning thermostat (e.g., Nest™) even maps your temporal behavior to record the temperature you desire at different times; it will also sense whether there is someone home and adjust accordingly. Someone jokingly explained that he thought a *thermos bottle* was intelligent because it knew that it should keep a cold beverage cold and a hot beverage warm.

In this acceptation of the term, intelligent behavior does not require understanding. Instead, learning means refining the map from situation to action. This map can be a table that the machine looks up. If this is Wednesday at 8:00pm, set the temperature to 72 degrees. Don't ask why, just do it.

I said that understanding means fitting a model to observations, like the motion of planets to Newton's laws. Discovering such laws is very powerful because the model can then apply to different situations. For instance, Newton's laws explain the motion of the Earth around the Sun, but also explain what happens when there are more planets. A machine learning approach could predict the position of the Earth around the Sun at a future time from many past observations. It would not be able to infer what would happen in a system with more planets. Understanding is more powerful, but is feasible only when a few laws suffice to fit the observations. What laws would explain the differences between cats and dogs? Interestingly, a child learns to distinguish dogs and cats after being exposed to very few examples, a feat that puts the best deep neural network to shame.

An important property of machine learning is 'generalization'. After training, the machine should be able to make correct decisions in future situations that have not been observed before. For instance, one shows thousands of pictures of dogs and cats to the machine and tells it which are dogs and which are cats. You then show a picture of a new dog, and the machine should guess correctly. To test if this works, one tries many new pictures and counts how often the machine makes mistakes. One cannot guarantee that the machine will be correct on any given future picture. In fact, a future picture may have characteristics that differentiate it from the previous ones. This may not matter too much when classifying pictures of cats and dogs, but it certainly does when trying to detect a cancer in an X-ray or a pedestrian in a camera image.

 ## Paradigm Shift

Engineers have been designing systems that 'learn' for decades. For instance, the telephone system uses filters first developed in the early 1960s that minimize the distortion that transmission lines introduce. These filters tune themselves automatically, thus 'learning' to minimize the distortion. The tuning is necessary as the distortion tends to vary significantly from call to call. Such a filter uses a model of the received signal as a superposition of delayed and attenuated versions of the transmitted signal. This model is based on the actual physics of the system.

As another example, assume you are observing the position of the Earth around the Sun on ten consecutive days and that you want to predict its position on the next day. Using Newton's laws, you can write the position of the Earth on successive days along an ellipse with nine parameters. From the ten observations, you can infer the parameters of the orbit and compute the next position. We call this methodology a 'white box' approach because it uses a model based on the physics of the actual system. The benefits of this approach are that the parametric model is as simple as possible and its only unknowns are a few parameters.

An alternative methodology is a 'black box' approach. When using this approach, one chooses a generic model of the system that does not reflect the actual 'physics' of the situation. In the case of the motion of the Earth, one might choose a model where the next position is a polynomial of degree two of the previous ten positions. Specifying this polynomial requires many parameters (about 1500 in three dimensions) and computing these parameters requires a very large number of observations. Thus, learning is inherently much slower when using a generic model instead of a compact physics-informed model.

The power of 'physical' models is even more striking when they can predict phenomena that have never been observed. Think of the bending of light by massive objects predicted by General Relativity and the existence of the Charm quark or of the Higgs boson predicted by particle physics. These predictions suggest the experiments to be performed to confirm the theory. This magic is outside of the capabilities of big data analysis! Watson – IBM's artificial brain – cannot be Einstein.

In some cases, there is no known model and a white box approach is not feasible. For instance, say that you want to build a system that determines whether a picture is that of a dog or a cat. You choose a parametric function that maps the picture X into a number Y between 0 and 1 that represents the probability that the picture is that of a dog. That is, one writes $Y = f(\theta, X)$ where $f(\cdot)$ is the function and θ are the parameters. One then 'trains' the system by showing it a number of pictures of dogs and cats. The training method is simple: say that the current parameters are θ and that you can calculate the gradient of $f(\theta, X)$ as G. The gradient indicates the direction in which one should modify θ to increase Y. Say that the next picture X is that of a dog and that the systems computes Y. We then adjust θ in the direction G that increases Y, to reinforce the decision that the picture shows a dog. If the next picture is that of a cat, we change θ in the direction that decreases Y, i.e., in the direction $-G$. After repeating this procedure many times, the system has 'learned' parameters θ. This method is called the *Stochastic Gradient Descent* algorithm. One then tests how well the system performs on a number of additional pictures. If one is satisfied with the results, one assumes that the resulting function is good enough. This function has no 'understanding' of what differentiates a dog from a cat. It does not constitute a 'physical' model.

These brute-force black box approaches are feasible for some problems because of the access to a large amount of data and to powerful computers or dedicated processors. They met with very little success when they were invented about seventy years ago because the technology at that time did not enable the scale required for them to be effective.

Deep Neural Networks

Deep Neural Networks (DNN) are electronic processing circuits inspired by the structure of the brain. Our vision system consists of layers. The first layer is in the retina that captures the intensity and color of zones in our field of vision. The next layer extracts edges and motion. The brain receives these signals and extracts higher level features. A model of this processing is that neurons are arranged in successive layers, where each neuron in one layer gets inputs from neurons in the previous layer through connections called synapses. Presumably, the weights of these connections get tuned as we grow up and learn to perform tasks, possibly by trial and error.

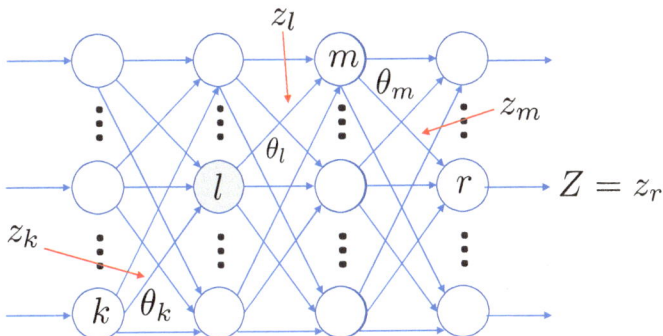

The figure sketches a DNN. The inputs at the left of the DNN are the features X from which the system produces the probability that X corresponds to a dog, or the estimate of some quantity.

Each circle is a circuit that we call a neuron. In the figure, z_k is the output of neuron k. It is multiplied by θ_k to contribute the quantity $\theta_k z_k$ to the total input V_l of neuron l. The parameter θ_k represents the strength of the connection between neuron k and neuron l. Thus, V_l is the sum of $\theta_n z_n$ over all the neurons n of the layer to the immediate left of neuron l, including neuron k. The output z_l of neuron l is equal to $f(a_l, V_l)$ where a_l is a parameter specific to that neuron and f is some function that we discuss later.

With this structure, it is easy to compute the derivative of some output Z with respect to some weight, say θ_k. That is, one can compute the gradient of the output with respect to the parameters. One then uses this gradient to implement the stochastic gradient descent algorithm we discussed earlier.

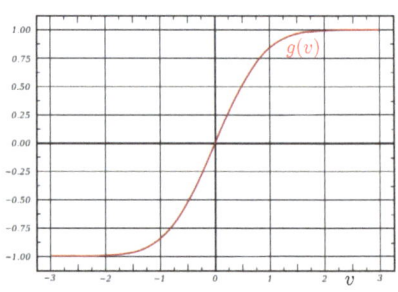

What should be the functions $f(a, V)$? Inspired by the idea that a neuron fires if it is excited enough, one may use a function $f(a, V)$ that is close to 1 if $V > a$ and close to -1 if $V < a$. A possible choice is $f(a, V) = g(V - a)$ with

$$g(v) = \frac{2}{1 + e^{-\beta v}} - 1$$

where β is a positive constant. If β is large, then $e^{-\beta v}$ goes from a very large to a very small value when v goes from negative to positive. Consequently, $g(v)$ goes from -1 to $+1$. This function $g(\cdot)$ is called a *logistic function*.

The DNN is able to model many functions by adjusting its parameters. To see why, consider neuron l. The output of this neuron indicates whether the linear combination $V_l = \sum_n \theta_n z_n$ is larger or smaller than the thresholds a_l of the neurons. Consequently, the first layer divides the set of inputs into regions separated by hyperplanes. The next layer then further divides these regions. The number of regions that can be obtained by this process is exponential in the number of layers. The final layer then assigns values to the regions, thus approximating a complex function of the input vector by an almost piecewise constant function.

The missing piece of the puzzle is that, unfortunately, generally the cost function is not a nice convex (bowl-shaped) function of the parameters of the DNN. Instead, it typically has many local minima. By using the stochastic gradient descent algorithm, the tuning of the DNN may get stuck in a local minimum. Also, to reduce the number of parameters to tune, one usually selects a few layers with fixed parameters, such as edge detectors in vision systems. Thus, the selection of the DNN becomes somewhat of an art, like cooking.

Thus, it remains impossible to predict whether the DNN will be a good technique for machine learning in a specific application. The answer of the practitioners it to try and see. If it works, they publish a paper or they implement and commercialize the system. We are far from the proven convergence results of specific

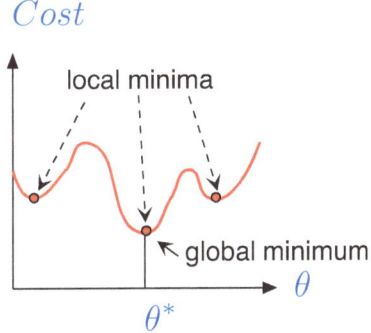

classes of systems such as echo cancellers where the physics-based model results in a convex optimization problem. Ah, nostalgia....

A Word of Caution

Black box approaches present a real risk. After training, the system has learned parameters of a function. Since the function does not correspond to a physical model, there is no real way to predict if the decision will be correct on a future input. Through experiments one can determine if the decision of the system changes dramatically when one modifies the input slightly. This is useful to test the sensitivity of the system to image distortions. However, since one does not know the possible differences in inputs that should yield the same output, this does not really predict the trustworthiness of the system.

Imagine then a killer drone with a DNN target recognition system .... It is not surprising that a number of serious scientists have raised concerns about 'artificial stupidity' and the need to build safeguards into such systems. "Open the pod bay doors, Hal."

Many other issues are relevant for 'artificially intelligent' systems. When presented with difficult choices, humans use a sense of morality. Should the self-driving car kill a baby or an eighty-year old, if these are the only two options? What confidence does the killing drone need before it hits a target? Is it possible to formulate moral principles in a way that one can build them into the machine? Can one teach morality to a machine?

 ## Bandit Problems

.

Bandit problems capture an interesting class of decision-making problems in an uncertain environment. You have two coins A and B, each with an unknown bias. Your goal is, at each step, to flip one of the two coins to accumulate as many 'heads' as possible.

After a number of steps, having flipped each coin a number of times, you see that A produced 'heads' 54% and B 57% of the times. Should you then flip B and keep on flipping it until its fraction of 'heads' falls below 54%? While this strategy may seem

A   B

reasonable, it is exposed to the risk that you may have been unlucky with coin A. Indeed, it could be that coin A actually yields 'heads' 62% of the times, but your flips of coin A were less successful that typical. If you stick to your strategy and coin B actually produces 'heads' 58% of the time, you might end up flipping coin B forever and miss out more and more on opportunities to accumulate 'heads'.

This simple example shows that you should never stick to a coin. Rather, you should keep on flipping both coins to protect yourself against the possibility of being stuck with the wrong coin. Thus, your should keep *exploring*. Obviously, you should tend to flip the more successful coin more frequently: you should *exploit* your knowledge of the coins.

What is then the best tradeoff between exploration and exploitation? Also, how well does the best strategy perform? Assume that, unknown to you, the bias of A is 60% and that of B is 56%. If you always flip coin A, after n coin flips, the expected number of 'heads' is $0.6n$. If you always flip coin B, it is $0.56n$, so that your expected 'loss' with respect to coin A is $0.04n$. In fact, any strategy that eventually gets stuck with coin B accumulates a loss approximately equal to $0.04n$ after a large number n of steps. We say that the expected *regret* grows linearly over time, i.e., as a multiple of time.

It turns out that the best possible strategy accumulates a regret that grows only like $\log(n)$ instead of linearly. This is cool because $\log(n)$ is much smaller than n for large n. For instance, when $n = 10^6, \log(n) = 6$, which is much smaller than n. Moreover a strategy that achieves this slow accumulation of regret is easy to describe: flip a coin according to its likelihood of being the best given past observations.

How can we figure out that likelihood given that we do not know anything about the coins? The trick is to assume some prior knowledge about the coins. For instance, let us assume that the bias of each coin is equally likely to be any value in $[0,1]$, uniformly. If we flip coin A n times and it pro-

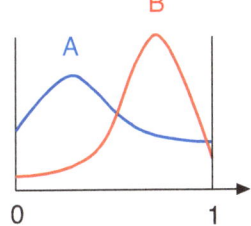

duces h 'heads', we can update the likelihood of the values of the bias of that coin. We do the same for coin B. At the next step, we then generate a random bias for coin A with the likelihood that we computed, and similarly for coin B. We then flip the coin that has the largest generated bias, and we continue. This algorithm, called *Thompson sampling* was proposed in 1933 and rediscovered less than ten years ago.

11

Information

I used to watch the local weather forecast on TV. (Remember TV?) In Berkeley, most summer days are foggy in the morning and evening, and sunny in mid-day. Eventually, I realized that there was no point in watching the forecast.

How much information is there in a message? Some mystery novels are fascinating because their plot is unpredictable. You keep on reading, hoping to guess 'who done it'. If the outcome is predictable, you lose interest. Clearly, there is information if the content of the message is unpredictable. Claude Shannon built on this intuition and developed a precise measure of the quantity of information in a message.

Coin Flips

Clause Shannon started with the simplest example: a coin flip. He defined the quantity of information in the outcome of a fair coin flip to be *one bit*. This makes sense: you need one bit 0 or 1 to describe the outcome 'tails' or 'heads' of the flip. If you flip the coin 100 times, you need 100 bits to describe the outcomes. Note that there are 2^{100} possible strings of 100 bits 0 and 1. Indeed, there are two possible choices for the first bit, then two for the second bit, and so on until the last bit. Thus, there are $2 \times 2 \times$

$\cdots \times 2 = 2^{100}$ possible strings. All these strings have the same probability.

Now assume that your coin is loaded and produces 'heads' only 10% of the time, instead of 50%. Intuitively, the outcomes of 100 coin flips are now more predictable: most of them are 'tails'. Accordingly, it should take less information to describe these outcomes. However, at first glance, you still need to describe the outcomes of the 100 coin flips, and that seems to require 100 bits. How can we get away with fewer bits?

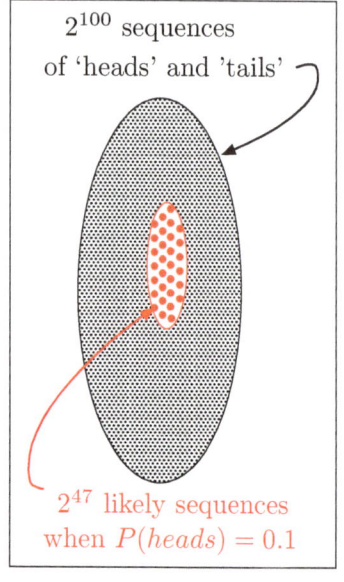

Shannon showed that there are not too many sequences of 100 coin flips with about 10% of 'heads'. There are only about 2^{47} such sequences and it turns out that they are equally likely to be produced by the coin flips. That's quite many, but much fewer than 2^{100}. We can number these sequences with strings of 47 bits, because there are 2^{47} such strings. Accordingly, we need only 47 bits to describe which of the 2^{47} strings was produced by the 100 coin flips. Hence, on average, we need 0.47 bit per coin flip to describe the outcomes, when the coin is biased and produces 'heads' 10% of the time.

More generally, the number of bits per coin flip is a function $H(p)$ of the bias p of the coin, i.e., the fraction p of the time that the flip produces 'heads'. This quantity $H(p)$ is called the *Shannon entropy*. In particular, $H(0.5) = 1$ since we need one bit per flip when the coin is fair. Also, we just saw that $H(0.1) = 0.47$. The

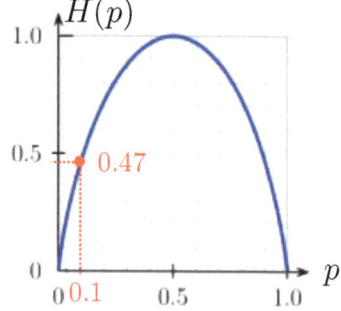

figure shows a plot of the function $H(p)$. Observe that $H(0) = 0$. This makes sense since, when $p = 0$, all the outcomes are 'tails' and it does not take any information to describe them.

 ## Book

Coin flips are fun, but how would you define the quantity of information in a book? This seems to be quite a formidable task. Also, even if you manage to do it, how useful can that possibly be?

The idea is similar as for coin flips. The number of sentences with 100 characters that you find in a book is much smaller than the possible number of such strings. This is not surprising. Many strings of characters (e.g., aegswa) never show up in a book. (By the way, I googled it and aegswa shows up in Luke 9, 52, but never mind.)

The Lempel-Ziv-Welch (LZW) coding scheme builds progressively a list of the longest strings it finds in the book. Along the way, when it sees again one of the strings in the list, it replaces it by its index in the list. The decoder builds the same list and recover the text from the indices. Since the number of strings is much smaller than the set of all possible strings, long strings are replaced by a small index, thereby reducing the number of bits needed to encode the book. It turns out that this scheme essentially achieves Shannon's entropy of the book, defined as the minimum number of bits per symbol required to encode the book.

Say what?

 ## Noisy Channel

It is difficult to be heard in a noisy restaurant. One has to talk loudly and slowly, and possibly add a few extra words to be understood without ambiguity. The same is true for electronic communication. For instance, the transmission rate over a noisy wireless channel is much slower than across an optical fiber.

For many years, communication engineers designed sophisticated schemes to improve the transmission rates. Unfortunately, they had no clue as to whether significant further improvements were still possible. The situation changed in 1948 when Shannon (yes, again) discovered that every communication channel has an absolute speed limit and he explained how to calculate that limit.

Shannon's result states that, for a given channel, it is possible to transmit bits reliably at a rate arbitrarily close to C, but not faster. Here, C is some number of bits per second called the *Shannon capacity* of the channel.

The precise meaning of this statement is subtle. If a channel is noisy, any given transmission can be corrupted. So, what does it mean to transmit reliably at some rate? It cannot mean to transmit without any error, since these are unavoidable. It means, roughly, with an arbitrarily small probability of error per bit.

Let's try to be precise. Let $R < C$. Shannon says that, for any given $\epsilon > 0$, there is some N large enough so that it is possible to transmit $R \times N$ bits in N time units with a probability of error less than ϵ. Moreover, this is not possible if $R > C$. That is, if $R > C$, there is no transmission scheme that achieves an arbitrarily small probability of error. Thus, 'reliably' means with an arbitrarily small probability of error. Note that this is possible if we transmit a large enough number of bits. (Technically, the value of N depends on ϵ: a smaller ϵ requires a larger N.) If you transmit any fixed number of bits, there is some positive probability that these bits get corrupted.

The details of this result are slightly complicated. However, the main idea is simple. It is that if you transmit a string x of 1000 bits, the number of possible error patterns that can corrupt x is relatively small. For instance, if the errors correspond to 'heads' in coin flips with $P(heads) = 10\%$, then there are only 2^{47}

likely error strings. Thus, the string x is corrupted into any one of 2^{47} strings. If we choose the different strings x that we transmit so that these sets of possible corrupted strings are distinct, then the receiver can figure out the string that was sent. Since each transmitted string can become any one of a set of 2^{47} corrupted strings, there can be at most $2^{100}/2^{47} = 2^{53}$ different strings x that result in disjoint sets of corrupted strings.

A careful analysis shows that this number is achievable. Surprisingly, the 2^{53} strings of 100 bits that one sends do not have to be selected with great care to guarantee that the resulting sets of corrupted strings are disjoint. They can be chosen by flipping coins!

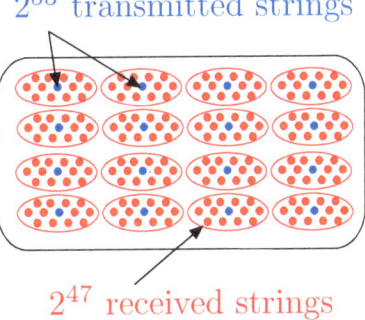

2^{53} transmitted strings

2^{47} received strings per transmitted string

Thus, one should be able to transmit any one of 2^{53} different strings in 100 time units. This means that the capacity of the channel C is equal to $C = 0.53$ bits per unit time.

Note that $C = 1 - H(p)$. Intuitively, this means that the capacity of a noisier channel is smaller.

12

The Arrow of Time

> *People like us, who believe in physics, know that the distinction made between past, present and future is nothing more than a persistent, stubborn illusion.*
>
> Albert Einstein

Here I address a question that keeps puzzling scientists and philosophers (and me). Why does time flow only forward even though the laws of physics can be reversed in time? One standard answer is that the 'randomness' of the world is increasing. I discuss that answer and the problems it poses. I then explain a different interpretation.

The entropy story

The standard answer for the irreversibility of time (Boltzmann, Eddington) is that, globally, the world is becoming more and more disordered. This phenomenon is familiar. For instance, when you pour some milk into your coffee, the two liquids mix and never separate again. This is surprising since the trajectory of any particular molecule of liquid is equally plausible when you reverse it in time. Similarly, when you start a billiards game by breaking the cluster of balls, these spread and will not come together again in their neat initial packed triangle formation. However, if you watch any single ball, you can imagine it going backwards along the same trajectory.

In order to examine the disorder story, let us look at a simple example. Imagine a closed box. (See the figure.) We divide the box into two halves, say left and right, with some membrane. We then inject 1,000 molecules of gas in the left-hand part of the box. After a while, we remove the membrane.

The molecules spread into the box and you don't expect to see them all again in the left side of the box. Thus, if you watch a movie of this experiment, you can tell whether it is played forward or backward.

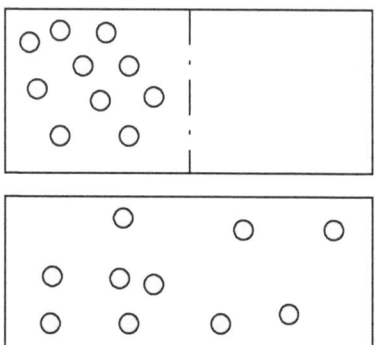

One can formulate this notion of randomness by using the entropy we discussed in the previous chapter. Consider a single molecule and simplify its description by saying that it has two possible states: either 0, which means that it is in the left-hand side of the box, or 1, which means that it is in the right-hand side. Initially, the state is 0 since the molecule starts in the left-hand side. Thus, initially, the probability that the state is 1 is equal to 0. After we remove the membrane, that molecule is free to move to the right-hand side, so that it states jumps from 0 to 1. A simple random model of the motion of the molecule shows that the probability that its state is 1 increases monotically over time. After a long time, the probability that the state is 1 approaches 1/2, because the molecule is equally likely to be in the left or right-hand side of the box, by symmetry.

At any given time, we can think of the state of the molecule as the outcome of a coin flip that has some probability p of 'heads'. As we saw in the previous chapter, this probability p is associated with a measure of entropy $H(p)$. Now, consider the 1000 molecules. They move randomly and independently. Thus, we can think of their states, at some given time, as corresponding to the outcomes of 1000 flips of coins that have bias p. As Shannon told us, there are about $2^{1000H(p)}$ equally likely strings of outcomes of these flips, i.e., of states of the molecules.

Boltzmann defined the entropy of these random states as the logarithm of the number of possible configurations. Thus, the Boltzmann entropy is $1000H(p)$, which agrees with Shannon's definition of entropy, even though they came to this notion from very different viewpoints.

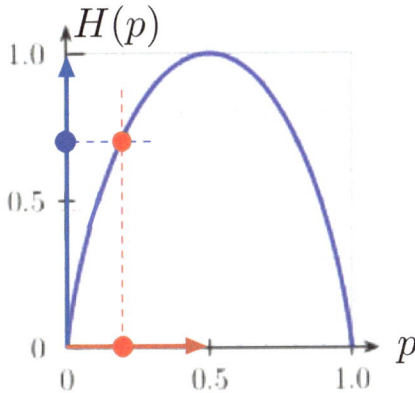

As p increases monotonically from 0 to 0.5, we see that the entropy $1000H(p)$ of the system increases monotonically from 0 to 1000. This increase of entropy proves the existence of an arrow of time, as Boltzmann and Eddington stated.

 ## Objection

A key objection to the entropy story is that entropy is not a physical quantity: it cannot be measured or 'sensed'.

The entropy is a mathematical quantity that depends on the likelihood of the different states that the world *could have been in*. At any given time, the world is in one specific state. A quantity that involves other possible states that it might have been in is not measurable. There is no measuring instrument that will respond to the entropy. The output of a measuring instrument can only be a function of the state of the world, or of its different states in the recent past.

Think about how our brain perceives time. This perception

can only be based on a measurable physical quantity, not on the likelihood of states that the world is not in. I believe that the arrow of time must involve *an irreversible measurable quantity*. So, entropy is not adequate. In the remainder of the chapter, I discuss such a possible irreversible measurable quantity.

 ## Function of state

Let us go back to our box with the gas molecules. Let F be the fraction of molecules in the right-hand side of the box. This is a measurable quantity. It corresponds to the density of molecules in the right-hand side of the box. One could construct a sensor, optical or chemical, that would measure that concentration. One may also imagine that our brain responds to such changes in concentration.

Initially, $F = 0$. At some later time, the probability that any one of the molecules is in the right-hand side of the box is p. At that time, we expect F to be close to p. In fact, in chapter 3, we saw that $F = p \pm 3\%$, with 95% confidence. Hence, one may expect that, as time increases, the quantity F will track p and roughly increase from 0 to about 0.5.

Could this increase of a measurable quantity be an arrow of time? Interestingly, the plot thickens, as the next section explains.

 ## The puzzle of recurrence

Many of you probably have realized that, at some time in the future, all the molecules will find themselves again in the left-hand side of the box. This will take a long time, but it will happen, inevitably. Consequently, the measured quantity F will at some future time be equal to zero again. This is called the *recurrence puzzle*.

I hear you say: Wait a minute! You told me that the entropy increases monotonically and now you observe that all the molecules will be in state 0 at some future time; that configuration has no disorder. Isn't that a contradiction? It is not. Remember that the entropy concerns probabilities of the states. Even states with small probabilities can occur.

To me, this very different behavior of the entropy and of measurable quantities is another argument against viewing entropy as an arrow of time.

You will certainly point out that since F returns to zero, it cannot be an arrow of time. I beg to differ, as I explain below.

 ## Infinite box

Some physicists who are uneasy about the entropy and the puzzle of recurrence have proposed that the universe is in fact an infinite box, so that the molecules do not all come back to the section where they started from.

In this view, the universe starts with the Big Bang, and expands forever. The story is subtle, because the expansion of the universe implies that the speed of objects increases in proportion to their distance. Thus, objects that are far enough move away from us faster than the speed of light and cannot be visible, so that we see only a finite universe. To make up for this, physicists invoke the eventual decrease in the energy density of the vacuum. (This constant is the proportionality factor between the distance and the expansion speed.) This argument is ingenious, but I think it is misguided.

Our perception of time cannot depend on the value of the energy density of the vacuum in a few billion years. It must be a local (in space and time) phenomenon.

Perception of time

Our subjective perception of time is due to a succession of events, like a movie is a sequence of pictures. We all perceive time differently. But why do we feel that we are 'now' and why do we remember yesterday and not tomorrow?

Say that we watch a single billiards ball. We see it moving from left to right for five seconds. This motion is reversible, according to the laws of physics. However, for us, it moves from left to right, not from right to left. We can *imagine* that the ball could have moved from right to left, but we do not recall having seen it to the right of its current position; we remember having seen it to the left. Thus, even though we observe a perfectly time-reversible motion, we distinguish the past and the future.

This asymmetry is not due to the increasing disorder of the universe outside of us, and certainly not to the increase of entropy of the ball we are watching. I believe it is due to the fact that our brain is a forward-only camera. The movie that it captures cannot be played backward.

The forward-only nature of our brain is due to the complexity of the brain processes, not to the outside world. I am not a neuroscientist, so my view of the brain mechanisms is very simplified. I imagine a neuron firing and sending a collection of neurotransmitter molecules through a synapse. These then excite other neurons, and so on. Picture this process as spilling beans that collect on some receptors. This elementary process corresponds to an increase in concentration of density of neurotransmitters at the receptors. This is a measurable irreversible effect, like the fraction of balls in the right-hand side of our box.

Short term versus long term

As we noted, our measured quantity F is recurrent. It starts from 0 and tends to follow p that increases from 0 to 0.5. However, at some future time, F is again equal to zero, and the process repeats forever. Not much of an arrow of time!

My contention is that F is a *local* (in time) arrow of time. It provides us with a sense of irreversibility of time that is convincing enough for our lifetime. Phenomena like our box are irreversible as they are perceived during the lifetime of humans and animals. I will elaborate on that point shortly.

Could there be beings whose perception is not governed by locally irreversible phenomena? For instance, would some being whose lifetime is longer than that the recurrence time of the world around it have a reversible view of time? Hard to know!

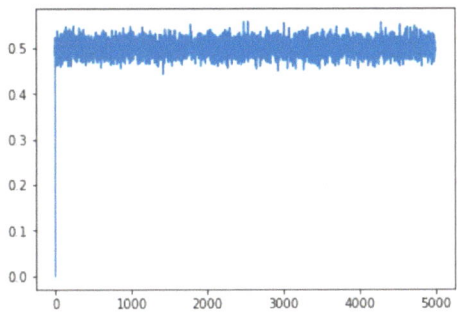

Does the irreversibility of F last long enough? The brain processes involve cascades of firings by neurons. If each firing appears irreversible during its own lifetime, then so will the cascade. Neuroscientists tell us that the relevant time scale for one firing is a few milliseconds. The figure shows F as a function of time. It starts from 0, then increases to about 0.5 and fluctuates. A simple model shows that the return time of F to zero is much larger than milliseconds when many molecules are involved. Moreover, significant decreases of F do not occur in milliseconds. Thus, it is plausible that these phenomena appear irreversible during their relevant time scale.

Let me restate the main point of this section. In our model, the irreversibility does not result from the increase of entropy. Instead, it arises from the temporary irreversibility of a function of the actual state. Unlike entropy, this function can be measured or 'sensed' and could explain the irreversibility of our perceived time.

Arrow of time in the brain

Summarizing, we argued that entropy is not credible as an arrow of time because entropy is a quantity that cannot be 'sensed'. (Sorry Boltzmann and Eddington!) Instead, we looked for a measurable irreversible quantity. In the brain, we can picture the concentrations of neurotransmitters near receptors as being such irreversible effects. Thus, the arrow of time is in our brain. This view is probably in agreement with Einstein's quote at the start of the chapter.

The arrow of time is in the brain, not outside

 ## How we see other phenomena

If we could watch the Earth moving around the Sun, we would see it tracing an ellipse in a specific direction. We would not remember it being further than it is, but we could remember having seen it in previous locations.

What about observing irreversible phenomena? Say you pour milk into your coffee. Why does our brain's sense of 'forward' time correspond to a mixing of milk and coffee and not to the separation of milk and coffee? In other words, why is our forward time consistent with the increasing disorder of the external world?

Let us consider a simple version of this question. Say we have two copies of our box. We perform the same experiment in parallel with the two boxes: inject gas in the left-hand side, remove the membrane, and see what happens. The same process occurs in the two boxes. If one box is a caricature of the processes in our brain, its forward time agrees with the increased disorder (locally in time) of the other box.

Thus, although the arrow of time is only in our brain, it imposes a unidirectional view of time on the outside world, even when we observe perfectly time-reversible phenomena. Hence, even if a device can see the future because it does not have an arrow of time, we are not be able to access future observations.

Our life is a trajectory in space-time. Seen globally, this trajectory could be traversed in both time directions. However, our brain takes the pictures only in one of these two directions.

13
Comments

The ideas in this book are not original. They can be found in one form or another in the references below and in many others. I apologize to the many authors whose work is not properly mentioned.

 ## Uncertainty

I don't know what is worse. A scientist dabbling in philosophy or a philosopher dabbling in science? This is why I am quite uneasy about this chapter. Serious philosophers have spent decades reflecting on the issues of consciousness and free will and I dismiss some of their conclusions in three sentences. My goal was certainly not to offend anyone and I apologize if my thought process appears shallow and lacking in knowledge of many substantial works.

The first figure is from [Keller, 1986]. For a more detailed study of coin flips, see [Diaconis et al., 2007]. Berliner [Berliner, 1992] discusses the relationship between chaos and probability. Bell's theorem is in [Bell, 1964]. The second figure is from [Schneider, D.R., 2014].

The discussion of consciousness and free will is influenced by the ideas of Dennett [Dennett, 2017, 1991], Searle [Searle, 2004],

Penrose [Penrose, 1989], and Poincaré [Poincaré, 1914].   I also enjoyed the last book of Hawking [Hawking, 2018]. I think that consciousness is probably the deepest unsolved mystery in science and I am sorry if my conclusion is somewhat anticlimactic: we cannot measure consciousness from outside and, therefore, it cannot be analyzed scientifically. I find it surprising that many writers do not distinguish the functional model and the explanation of consciousness, which to me is fundamental (e.g., [Tononi and Koch, 2015]). For a discussion of what it is like to be a bat, see [Nagel, 1974].

The experiments on commitment before awareness are discussed in [Smith, 2011].

The cute graphic on morals is from [morals].

The key ideas of evolution are from Darwin [Darwin, 1859]. I am sad for anyone who does not marvel at the beauty of those ideas. The role of sex in evolution was formulated mathematically in [Livnat et al., 2008].   The tree of evolution can be explored at [onezoom].

The axioms of probability are explained in [Kolmogorov]. There are many excellent texts on the topic, but it is not necessary to explore them to understand this little book. For an insightful and inspiring discussion of the foundations of Probability Theory, I highly recommend [Fine, 1973].

 Risks

It is difficult to be rational about risks. Emotions get in the way. Figure 2.1 is based on statistics in [Mosher, D. and Gould, S.]. Related statistics can be found in many places, e.g., [Insurance Information Institute], [American Cancer Institute]. The abysmal effectiveness of TSA screening is discussed in [Kerley, D. and Cook, J.]. Taleb's books, such as [Taleb, 2004], are engaging discussions of the hidden role of randomness in life.

Causation

The confusion between correlation and causation is the topic of many writings. The graphs in [Vigen, T.] show amusing correlations that make it obvious that correlation does not imply causation. As an example, the number of worldwide non-commercial space launches is remarkably correlated with the number of sociology doctorates awarded in the United States.

In the examples, we choose 95% confidence intervals to determine if a circumstance (like smoking) and a symptom (such as lung cancer) are correlated. Statisticians say that the *P-value* of the test is then 5%.

The popular misinterpretation of confidence intervals is highlighted in [Amrhein and al., 3/20/2019] (thanks to Prof. Satish Rao for pointing out this article).

[Thiese, M.S., 2014] presents a nice discussion of observational and interventional studies.

Confidence intervals are essential to understand statistical significance. They are explained in *[Confidence intervals]*. You may wonder what happens if you test fewer or more that 1000 people. How does the 95% confidence interval change in that case? The answer is that if you sample $N \geq 100$ people, then the confidence interval is F plus minus $1/\sqrt{N}$. Thus, for $N = 100$, it is $F \pm 10\%$. For $N = 400$, it is $F \pm 5\%$. Also, for $N = 1,600$, it is $F \pm 2.5\%$.

Simpson's paradox is analyzed in [Simpson, 1951] but had been observed previously by statisticians. The misleading aspects of statistics are described with humor in [Huff, 1954]. The excellent book [Kahneman, 2011] explains many ways in which our thinking is distorted.

Inference

The hearts and spades example illustrates a famous calculation known as Bayes' Rule (see e.g., [Bayes]). The Ebola or flu example is another illustration. Roughly, the problem is to guess the condition of a symptom. One knows the likelihood that each condition is accompanied by the symptom and also the prevalence of the possible conditions. To find the likelihood of the condition given the symptom, one calculates the fraction of individuals with the observed symptom and the condition among the individuals with the symptom.

 ## Gambling

The chapter starts with a discussion of how we feel about uncertain outcomes. The book [Kahneman, 2011] discusses the idea of utility maximization and explains *prospect theory* that is a more accurate model of how we tend to make decisions. Experiments show that we do not like a gamble where we are equally likely to win or lose $1000.00. We are more sensitive to losses than to gains. Prospect theory models that effect and seems better at predicting behavior than a simple expected utility maximization.

For a mathematical study of games of chance, see [Dubins and Savage, 2014].

The Kelly strategy is introduced in [Kelly, 1956]. A nice description can be found on Wikipedia.

 ## Randomized Algorithms

This little chapter only scratches the surface of an exciting area.

There are many excellent books on randomized algorithms (see e.g., [Motwani and P.]. The generation of random numbers is explained in the classic book [Knuth].

The time complexity of algorithms refers to the number of steps it takes to complete, usually as a function of a quantity that describes the number of inputs. For many deterministic algorithms, one typically worries about the best and worst cases. See [Karp,

1986] for an inspiring presentation and [Papadimitriou, 1994] for a comprehensive discussion.

 ## Auctions

The second price auction was analyzed by William Vickrey (Nobel Prize in Economics in 1996). Roger Myerson (Nobel Prize in Economics in 2007) derived the first price auction that maximizes the revenue. For a comprehensive discussion of auctions, see [Krishna, 2010].

 ## Games

Games are situations that involve multiple players and where the reward of one player depends on the actions of the other players. Game theory was initiated by John von Neumann, Oskar Morgenstern, (see [von Neumann and Morgenstern, 1944] and John Nash [Nash, 1950]). See [Game] for a nice presentation of the history of the field.

Learning in games is a topic of current research. See [Fudenberg] for an accessible introduction.

 ## Planning

That chapter explains the dynamic programming approach for computing the best decisions. The key notions are those of *state* and *value of a state* at a given time. The state of a system allows to predict its future evolution independently of the past; it summarizes everything we need to know about the system at a given time. The value of a state at a given time is the minimum cost starting from that state at that time. Proceeding backwards in time, one can compute the value of the states at given times recursively by solving the dynamic programming equations. These ideas were developed by Richard Bellman [Bellman, 1953]. That monograph explains dynamic programming for deterministic and stochastic systems, and also for games.

The Principle of Least Action has an old history. See [Feynman] for a clear discussion. In particular, in the case of light finding the fastest path, Feynman explains that all the rays of light that travel along other paths cancel one another because they arrive out of phase; only the paths close to the fastest one add up.

 ## Machine Learning

Machine Learning is the name given to a collection of algorithms that involve tuning parameters of decision functions. Such algorithms include classification of objects using planes to separate vectors of features (e.g., support vector machines), fitting an estimation polynomial (e.g., linear and nonlinear regression), and fitting the parameters of a neural network (deep or shallow).

I like to joke that deep learning is neither, in the sense that it does not lead to deep understanding.

Neural networks were proposed in 1943 by Warren McCulloch and Walter Pitts [McCulloch and Pitts, 1943]. The stochastic gradient descent learning scheme was introduced by Bernard Widrow and Ted Hoff in 1960 and also by Robert Lucky in 1964.

Properties of such algorithms are well understood when the cost function is convex. However, in the non-convex case, much remains to be explored.

There are many books that present the key ideas of machine learning in an accessible way and others that offer hands-on examples. The text [Murphy, 2012] is particularly clear.

The analysis of Thompson sampling can be found in [Agrawal and N., 2012].

Information

The notions of information and channel capacity are explained in the remarkable paper [Shannon, 1948]. This paper is a must-read for anyone interested in data, information, and communication. The book [Lucky, 1989] is an entertaining discussion of information and its processing by humans and machines.

The Arrow of Time

In Chapter 3 of his fun 1928 book, Eddington stated that time is a property of entropy alone [Eddington].

For a more general description of the ambiguity of the entropy story, consider a continuous-time Markov chain. This is a simple model of random evolution. The Markov chain has a number of possible states and it jumps from one to the other in a memoryless way. If the Markov chain has finitely many states and if each state can be reached from every other, then it approaches a statistical equilibrium that corresponds to a likelihood of being in each state that does not change over time. Interestingly, a well-defined distance between the likelihood of the states and their equilibrium likelihood decreases monotonically, thus providing an 'Eddingtonian' arrow of time. However, the Markov chain keeps on visiting all the states, thus preventing the observer from seeing any arrow of time.

Many popular books repeat the entropy justification of the arrow of time, but they often fail to note the recurrence paradox and the fact that it does not explain our perception of time.

The possible connection between the arrow of time and the energy density of vacuum is from [Susskind].

For a brain-centric view of the arrow of time, see e.g., [Smith, 2014].

A quantum-mechanical discussion can be found in [Podolskiy and Lanza, 2016]. The authors postulate that the arrow of time results from the increase in quantum mutual information between the observer and the observed world, so that the observer remembers the past, not the future. This analysis involves the probability of states the world is not in, through the mutual information. Consequently, this explanation is subject to the same objection as the entropy story.

14
Bibliography

S. Agrawal and Goyal N. Analysis of Thompson Sampling for the multi-armed bandit problem. *Proceedings of the 25th Annual Conference on Learning Theory (COLT'12)*, 2012.

American Cancer Institute. Lifetime risk of developing or dying from cancer. `https://www.cancer.org/cancer/cancer-basics/lifetime-probability-of-developing-or-dying-from-cancer.html`, Last accessed on 2019-3-18.

V. Amrhein and al. Scientists rise up against statistical significance. *Nature*, 3/20/2019.

T. Bayes. Bayes' theorem. `https://en.wikipedia.org/wiki/Bayes'_theorem`, Last accessed on 2019-3-18.

J.S. Bell. On the Einstein Podolsky Rosen paradox. *Physics*, 1 (3): 195–290, 1964.

R. Bellman. *An Introduction to the Theory of Dynamic Programming*, volume R-245. Rand Report, 1953.

L.M. Berliner. Statistics, probability and chaos. *Statistical Science*, 7(1):69–122, 1992.

R. Darwin. *On the Origin of Species*. John Murray, London, 1859.

D. Dennett. *Consciousness Explained*. Litle, Brown and Company, 1991.

D. Dennett. *Brainstorms: Philosophical Essays on Mind and Psychology*. The MIT Press, 2017.

P. Diaconis, Holmes S., and Montgomery R. Dynamical bias in the coin toss. *SIAM Review*, 49(2):211–235, 2007.

L.E. Dubins and L.J. Savage. *How to Gamble If You Must: Inequalities for Stochastic Processes, 3rd Edition*. Dover, 2014.

A.S. Eddington. *The Nature of the Physical World*.

R. Feynman. The feynman lectures on physics, vol. ii, ch. 19: The principle of least action. http://www.feynmanlectures.caltech.edu.

T.L. Fine. *Theories of Probability: An Examination of Foundations*. Academic Press, 1973.

D. Fudenberg. Learning in games. https://simons.berkeley.edu/talks/learning-in-games, Last accessed on 2019-3-18.

Game. Game theory. https://en.wikipedia.org/wiki/Game_theory#cite_note-5, Last accessed on 2019-3-18.

S. Hawking. *Brief Answers to The Big Questions*. Penguin Random House, 2018.

D. Huff. *How to Lie with Statistics*. Norton & Company, 1954.

Insurance Information Institute. Facts + statistics: Mortality risks. https://www.iii.org/fact-statistic/facts-statistics-mortality-risk, Last accessed on 2019-3-18.

D. Kahneman. *Thinking, Fast and Slow*. Farrar, Straus and Giroux, 2011.

R.M. Karp. Combinatorics, complexity, and randomness: Turing award lecture on complexity and np-completeness. *ACM*, 1986.

J.B. Keller. The probability of heads. *Amer. Math. Monthly*, 93:191–197, 1986.

J.L. Kelly. A New Interpretation of Information Rate. *Bell System Technical Journal*, 35 (4):917–926, 1956.

Kerley, D. and Cook, J. Tsa fails most tests in latest undercover operation at us airports. https://abcnews.go.com/US/tsa-fails-tests-latest-undercover-operation-us-airports/story?id=51022188, Last accessed on 2017-11-10.

D. Knuth. *The Art of Computer Programming Vol. 2: Seminumerical Algorithms, 3rd Edition*.

Kolmogorov. Probability axioms. https://en.wikipedia.org/wiki/Probability_axioms, Last accessed on 2019-3-18.

V. Krishna. *Auction Theory, Second Edition*. Academic Press, 2010.

A. Livnat, C. Papadimitriou, J. Dushov, and Feldman M.W. A mixability theory for the role of sex in evolution. *PNAS*, 105 (50):19803–1980, 2008.

R.W. Lucky. *Silicon Dreams: Information, Man, and Machine*. St Martins Pr, 1989.

W. McCulloch and W. Pitts. A logical calculus of ideas immanent in nervous activity. *Bulletin of Mathematical Biophysics*, 5:115–133, 1943.

Mosher, D. and Gould, S. The odds that a gun will kill the average american may surprise you. https://www.businessinsider.com/us-gun-death-murder-risk-statistics-2018-3, Last accessed on 2019-3-17.

R. Motwani and Raghavan P. *Randomized Algorithms*.

K.P. Murphy. *Machine Learning: A Probabilistic Perspective*. MIT, 2012.

T. Nagel. What is it like to be a bat? *The Philosophical Review*, 83 (4):435–450, 1974.

J.F. Nash. Equilibrium points in *n*-person games. *PNAS*, pages 48–49, 1950.

onezoom. Tree of evolution. http://www.onezoom.org.

C.H. Papadimitriou. *Computational Complexity*. Addison-Wesley, 1994.

R. Penrose. *The Emperor's New Mind*. Oxford University Press, 1989.

D. Podolskiy and R. Lanza. On decoherence in quantum gravity. *Journal-ref: Annalen der Physik*, 528(9-10):663–676, 2016.

H. Poincaré. *Science and Method*. Dover Publications, Inc., 1914.

Schneider, D.R. Bell's theorem with easy math, 2014. https://drchinese.com/David/Bell_Theorem_Easy_Math.htm, Last accessed on 2019-3-18.

J. Searle. *Mind: A Brief Introduction*. Oxford University Press, 2004.

C.E. Shannon. A mathematical theory of communication. *The Bell System Technical Journal*, 277:623–656, 1948.

E.H. Simpson. The Interpretation of Interaction in Contingency Tables. *Journal of the Royal Statistical Society, Series B*, 13:238–241, 1951.

K. Smith. Neuroscience vs philosophy: Taking aim at free will. *Nature*, 477 (7362):23–25, 2011.

R. Smith. Do brains have an arrow of time? *Philosophy of Science*, 81(2):265–275, 2014.

L. Susskind. The birth of the universe and the origin of laws of physics - cornell messenger lecture, 4/2014. http://www.cornell.edu/video/leonard-susskind-1-boltzmann-and-the-arrow-of-time.

N.N. Taleb. *Fooled by Randomness*. Random House., 2004.

Thiese, M.S. Observational and interventional study design types; an overview, 2014. https://www.ncbi.nlm.nih.gov/pmc/articles/PMC4083571/.

G. Tononi and C. Koch. Consciousness: here, there and every-where? *Phil. Trans. R. Soc. B*, 370, 2015. http://dx.doi.org/10.1098/rstb.2014.0167.

Vigen, T. Spurious correlations. http://www.tylervigen.com/spurious-correlations, Last accessed on 2019-3-10.

J. von Neumann and O. Morgenstern. *On the Theory of Games of Strategy*. Princeton University Press, 1944.

Index

www.ingramcontent.com/pod-product-compliance
Lightning Source LLC
Chambersburg PA
CBHW041212180526
45172CB00016B/27